T0135749

BINAURAL TECHNOLOGY FOR VIRTUAL REALITY

Von der Fakulät für Elektrotechnik und Informationstechnik der
Rheinisch-Westfälischen Technischen Hochschule Aachen
zur Erlangung des akademischen Grades eines
DOKTORS DER INGENIEURWISSENSCHAFTEN
genehmigte Dissertation

vorgelegt von

Diplom-Ingenieur
Tobias Lentz
aus Rheydt

Berichter: Universitätsprofessor Dr. rer. nat. Michael Vorländer
Universitätsprofessor Christian Bischof, Ph.D.

Tag der mündlichen Prüfung: 20. November 2007

Diese Dissertation ist auf den Internetseiten der Hochschulbibliothek online verfügbar.

Tobias Lentz

Binaural Technology
for Virtual Reality

Logos Verlag Berlin GmbH

λογος

Aachener Beiträge zur Technischen Akustik

Herausgeber:
Prof. Dr. rer. nat. Michael Vorländer
Institut für Technische Akustik
RWTH Aachen
52056 Aachen
www.akustik.rwth-aachen.de

Bibliografische Information der Deutschen Nationalbibliothek

Die Deutsche Nationalbibliothek verzeichnet diese Publikation in der
Deutschen Nationalbibliografie; detaillierte bibliografische Daten sind
im Internet über http://dnb.d-nb.de abrufbar.

Dissertation RWTH Aachen
D 82, 2008

ISBN 978-3-8325-1935-3
ISSN 1866-3052
Band 6

Logos Verlag Berlin GmbH
Comeniushof, Gubener Str. 47,
10243 Berlin
Tel.:   +49 (0)30 / 42 85 10 90
Fax: +49 (0)30 / 42 85 10 92
http://www.logos-verlag.de

Contents

Abstract - Zusammenfassung

Abstract

The generation and use of non-intrusive artificial virtual environments is gaining more and more importance. Virtual environments are used in a variety of fields such as product design or evaluation of prototypes. Moreover they turned out to be very effective for the visualization of complex data sets. In the past, investigations were focused mainly on the visual reproduction technique to present geometrical data in a three-dimensional way (stereoscopic representation). However, the human perception consists not only of visual input but is based on a number of sensations and thus it would be worthwhile to create multi-modal and interactive virtual environments.

In this thesis, first of all the techniques required to include the acoustic component into a virtual environment are described and assessed. Furthermore the implementation of a software system is described, which takes advantage of these techniques to create complex acoustical scenes in real time. It features spatially distributed sound sources which are utilized to create an environment that is as authentic as possible.

The system is based on the binaural technology (binaural: "concerning both ears") and aims at reproducing a sound for the ears of the user, that is equivalent to the sound in an original surrounding. It is essential to set up a sound field that is as exact as possible, to achieve a simulation with the highest degree of authenticity. This comprises a description of the source, including its relevant angle-, distance- and time-dependent radiation, the sound distribution in the virtual scene (room acoustics), the perception-related consideration of all sound field components, as well as the exact reproduction of the artificial sound at the ears of the user.

Therefore, the focus of the thesis is also put on the reproduction technology. In this context, an approach for dynamic crosstalk cancellations is presented, which enables a loudspeaker-based reproduction for binaural acoustical imaging instead of using headphones. Filters are necessary to ensure the required channel separation in a dynamic setting. These filters are calculated in real time on the basis of the given data concerning the position and measured transfer functions of the outer ear. Furthermore the integration of this spatial audio system into a Virtual Reality display system (five-sided CAVE-like environment) at the Center for Computing and Communication, RWTH Aachen University, is described and evaluated.

Zusammenfassung

Die Erzeugung und Nutzung künstlicher virtueller Umgebungen gewinnt immer mehr an Bedeutung und wird vor allem in Bereichen wie dem Produktdesign, der Prototypen-Evaluierung und in der Forschung zur Veranschaulichung komplexer Datensätze eingesetzt. In der Vergangenheit lag der Schwerpunkt auf der visuellen Darstellung um beliebige Geometrien dreidimensional anzuzeigen (stereoskopische Darstellung). Da sich die Wahrnehmung jedoch aus einer Vielzahl verschiedener Sinneseindrücke zusammensetzt, ist es wünschenswert, auch die Repräsentation der virtuellen Szenen multi-modal und interaktiv zu gestalten.

Im Rahmen dieser Arbeit werden zunächst Techniken beschrieben und evaluiert, mit denen eine Erweiterung der Darstellung um die akustische Komponente möglich ist. Des Weiteren wird die Implementierung eines Softwaresystems beschrieben, welches die vorgestellten Techniken nutzt, um komplexe akustische Szenen mit räumlich verteilten Schallquellen möglichst authentisch in Echtzeit zu realisieren.

Das vorgestellte System verwendet die Binauraltechnik (binaural: „beidohrig") mit dem Ziel, an den Ohren des Benutzers das Schallsignal zu reproduzieren, das auch im Original-Umfeld dort herrschen würde. Um eine möglichst authentische Simulation zu gewährleisten, ist es erforderlich, alle beteiligten Komponenten, die das Schallfeld beeinflussen, mit einer höchstmöglichen Genauigkeit nachzubilden. Dazu gehört die Beschreibung der Quelle mit ihrem charakteristischen winkel-, abstands- und zeitabhängigen Abstrahlverhalten, die Schallausbreitung in der virtuellen Szene (Raumakustik), die gehörbezogene Berücksichtigung (binaural) aller Schallfeldanteile und letztendlich die exakte Reproduktion dieses künstlichen Schallsignals an den Ohren des Benutzers.

Ein besonderer Schwerpunkt wird in dieser Arbeit auf die Reproduktionstechnik gelegt. Es wird eine dynamische Übersprechkompensation vorgestellt, die eine Wiedergabe über Lautsprecher ermöglicht. Um die benötigte Kanaltrennung für eine korrekte binaurale Wiedergabe auch im dynamischen Fall zu garantieren, werden die benötigten Filter zur Laufzeit entsprechend den Positionsinformationen mit gemessenen Außenohrübertragungsfunktionen berechnet. Schließlich wird die Integration dieses Audiosystems in das am Rechen- und Kommunikationszentrum vorhandene fünfseitige VR-Displaysystem beschrieben und evaluiert.

Chapter 1

Introduction

Virtual Reality (VR) denotes an environment in which human senses will be addressed by a computer generated representation of a virtual scene. This representation has to be transmitted by adequate reproduction techniques to the user. Besides the stimulation of human senses as such, an interaction with the virtual scene should be possible. In that sense, the immersion of the user into the scene is of great importance and can be defined as addressing ideally all human sensory subsystems in a natural way. Since a perfect immersion is, however, not possible, the degree of immersion reflects the quality of the generated virtual environment. In this context, the degree of immersion depends on the number of human senses being addressed (visual, auditory, tactile, and olfactory sense) and the naturalness of the reproduction. The visual sense is considered most important with regard to human perception, and thus, the main focus in the field of VR was put on computer graphics. The most sophisticated results for a wide range of applications can be achieved with room-mounted projection systems generating holographic images for a high degree of visual immersion. In contrast to Head-Mounted Displays (HMDs), the advantage of this solution is, besides the extended field of view, the immense rise of comfort and naturalness of perception. The development was focused on the clear aim of minimizing attachments and encumbrances in order to improve user acceptance [CNSD93]. In that sense, much of the credibility that CAVE-like environments earned in recent years is due to the fact that they try to be totally non-intrusive VR systems. All further considerations in this thesis are based on the usage of room-mounted displays.

Common research in the field of VR considers acoustic stimulation as addition to the visual representation a highly important necessity for enhancing immersion into virtual scenes [NSG02]. To create a VR environment with a satisfactory congruent visual and auditory representation, a precise spatial audio reproduction system is required [Beg91]. As a consequence of the aim of being as non-intrusive as possible, a loudspeaker-based acoustical reproduction system seems to be the most desired solution for acoustical imaging instead of headphone reproduction [ALK06]. Users should be able to step into the virtual environment without too much preparation or calibration, but still be immersed in a plausible scenery.

On the condition of using loudspeakers for reproduction, several realizations for a spatial representation of sound are possible. The 5.1 surround technology, which is sufficient for many applications, is very common to place the virtual sound sources into the scene. But for some applications a more sophisticated solution is necessary, especially when a very precise sound image is required with regard to position and distance. A well-known technique for creating a more precise auditory representation is Wave-Field Synthesis (WFS) [BVdV92]. This technique uses a large number of loudspeakers (up to several hundreds) to recreate the entire wave field in the listening area. The main advantage is that more than one user can listen to the simulated sound field without tracking the individual positions. For a proper two-dimensional sound field synthesis with a reasonably high spatial aliasing frequency of about 5 kHz, loudspeakers must be placed nearly every 40 mm in a plane all around the listener. If true three-dimensional scenes are required, the efforts to be made will even be higher. The placement of the large loudspeaker arrays needed for WFS systems may be very problematic, especially for the environments focused on here. Most loudspeaker-based spatial audio systems for CAVE-like environments use Vector Base Amplitude Panning (VBAP) to generate the virtual sound sources. One solution can be found in [LHS99]. This system uses soft canvas for video projection (three walls and floor), which offers the possibility of placing loudspeakers behind the screens, outside the beams of the projectors.

If the placement of loudspeakers behind the projection screens is not possible due to rigid screens, alternative solutions will have to be be found. Figure 1.1 shows a picture of the *CAVE-like* environment, the spatial audio system described later in this thesis, has been designed for. Since the backside can be closed and the floor is also used as a video projection plane, there is only the ceiling which provides enough space for placing loudspeakers. In this environment it is not possible to use any WFS or VBAP systems because the loudspeakers cannot be placed below or at the height of the listener's head. This calls for an alternative method of bringing spatial

Figure 1.1: Picture of the *CAVE-like* environment at RWTH Aachen University.

sound information to the listener's ears. In such case, the complete binaural approach presented here has many advantages.

A binaural signal defines the sound pressure level at the ears of the listener and contains already all spatial cues needed to assign a tree-dimensional direction to a sound event. With an adequate filtering, called binaural synthesis, it is possible to add these spatial cues to an arbitrary mono signal to generate a virtual sound source in every direction. Furthermore, it is possible to realize near-to-head sources very realistically considering the relevant characteristics caused by sources close to the head. In addition to directional cues, the characteristic of the source radiation, the directivity, is important, especially for dynamic and interactive applications. If both the listener- as well as the source-related characteristics are considered to an adequate extent, the realization of a spatial sound will be possible with a high degree of realism and naturalness. Besides the direct sound path from the source to the listener reflections have to be considered in virtual scenes which are no pure free-field environments. The incidence of each reflection has a different direction which can also be solved using the binaural representation.

The prerequisite for a correct spatial perception is an appropriate reproduction of each channel related to the corresponding listener's ear. In case of loudspeaker reproduction, an additional filtering has to be applied to ensure a sufficient channel separation taking under consideration the transmission from each loudspeaker to each ear. The filter technique to suppress the crosstalk between the channels is called Crosstalk Cancellation (CTC). In addition, it is necessary that the filters can adapt to the listener's position and orientation. A dynamic binaural synthesis in combination with a dynamic CTC, on the other hand, is an interesting way of allowing spatial auditory representation by using a small number of loudspeakers. In this case, however, only a single person can act in the virtual scene because a precise reproduction of the binaural signals is only possible at designated points in space (i.e. at the listener's ears). Furthermore, tracking the listener's position is also needed to adjust the filter set which allows the movement of the user. This solution is absolutely appropriate in VR environments such as a *CAVE-like* display or a *L-Bench* where the stereoscopic video imaging technique already requires the use of a head tracker. In most existing *CAVE-like* environments, as well as in the environment for which the spatial audio system has been designed, creating three-dimensional images is not possible for more than one user.

1.1   Organization of document

This thesis is organized as follows:
The chronological order of the chapters is based on the transmission of the sound from a source to the listener. First of all, Chapter 2 *Virtual Sources* describes the characteristic of the source which has to be modeled in the virtual scene. A special focus is put on the generation of directivity data for a realistic simulation of sources with non-uniform sound radiation, as i.e. natural instruments. The spatial information needed for a correct localization is added by transforming the single signal which is representing the source into a binaural representation which is related to the signals at the ears of the listener. The synthesis of a binaural signal including the particular characteristics of near-field Head-Related Transfer Functions (HRTFs) will be discussed in Chapter 3 *Binaural Synthesis*. With the synthesis and the source directivity it is possible to generate an audio signal which contains all spatial information of the virtual sources.

The CTC required for a correct reproduction of these binaural signals will be discussed in Chapter 4 *Crosstalk Cancellation*. Starting with an overview of the theory

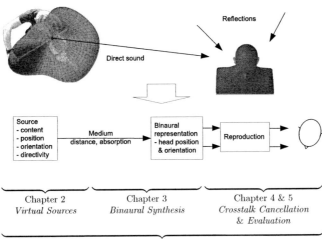

Chapter 2   Chapter 3   Chapter 4 & 5
Virtual Sources   *Binaural Synthesis*   *Crosstalk Cancellation*
                                           & Evaluation

Chapter 6 & 7   *Interactive VR-System* & *Validation*

Figure 1.2: Transferring the natural perception of a sound source into several steps of the simulation.

for a static solution and an investigation of the *sweet spot* which represents the valid area of the compensation, the dynamic solution will be introduced which has the ability to adapt on the listener's current position. A special focus is put on the stability of the generated filters and the usability without restrictions to the position of the listener. Chapter 5 *Evaluation* completes the description of the crosstalk cancellation with the evaluation of the channel separation achieved by the dynamic system.

Given the theoretical and technical background, Chapter 6 *Interactive VR-System* focuses on the integration of the technologies described into a software-based system for reproducing spatial distributed sound sources in the VR environment. After a brief overview of the visual reproduction system, special requirements will be assessed and the implementations resulting therefrom will be discussed. The connection to the VR toolkit ViSTA (Virtual Reality for Scientific Technical Applications) necessary for generating complex virtual audio visual scenarios is described and the performance achieved of the complete system will be reviewed. Furthermore the consideration of reflections calculated by an additional room acoustical simulation toolkit, the extended binaural filter calculation and fast convolution is introduced.

Finally, the localization performance which is achieved using the complete VR system is verified and discussed in Chapter 7 *Validation*.

It should be kept in mind that all considerations on each special issue are made on the background of the later use in the dynamic working VR system which requires that the used techniques have the ability to adapt on the changing environmental constrains, such as source positions listener position and the environment itself.

Chapter 2

Virtual Sources

When describing the transfer function from the source to the drain (listener), the directivity of the source and the distance need to be considered for a complete representation. Especially in an interactive system, where the user is able to displace the virtual source, or where the source itself is moving, the consideration of the radiation and the distance are additional cues for enhancing the plausibility of the complete environment. For a physically correct description it is not valid to separate the examinations concerning the directivity, the distance, and furthermore, the drain itself. Nevertheless, the technical constrains require a separation as a complete closed simulation of radiation and reception of an individual source and drain is not possible in real-time (up to now).

2.1 Source Directivity

A simple approach is to define directivity cones representing regions of a constant level. Sources with a simple directivity can be simulated by this method. However, often a more exact representation is desired, especially for natural instruments with a rather complex and, above all, a frequency-dependent directivity. Particularly for natural instruments the directivity determination in the complete sphere is not as simple as for electro-acoustical sources. The set of transfer functions defining the directivity of an electro-acoustical source can be measured consecutively with only one microphone and an appropriate deterministic excitation signal in any spatial resolution (see also the description of the measurement loudspeaker in Section 3.2). Contrary thereto, the possibility of sequential measurements in different directions is eliminated by the excitation variability of a human player. In this case, signals of all directions have to be recorded simultaneously to obtain comparable results with regard to a congruent excitation. Due to the fact that the number of microphones is limited (a spatial resolution of 5 degree requires $2,522$ microphones), an appropriate interpolation method is required.

In this section the measurements of various natural instruments are introduced. The influence of averaging the directivity information extracted from different single tones of a scale is analyzed and compared to the directivity extracted from a short piece of music [Sle04, LS06]. It should be mentioned, that in this context directivity is interpreted more precisely as relative directivity. This means that each transfer function of any direction is a relative transfer function being related to one specific direction, which is referred to as the reference direction. In this direction, the transfer function is ideally flat. It has to be ensured that the audio signal used for the playback was recorded with a microphone in the same direction as the reference direction of the directivity. Otherwise the directivity has to be recalculated with the new reference direction. The frequency characteristic of the recorded instrument at the reference direction, together with the relative or normalized directivity, leads to the complete specification of the instrument in any direction.

2.1.1   Measurement

All instruments were recorded in the full-anechoic chamber (interior dimensions $5\,\mathrm{m} \times 4\,\mathrm{m} \times 3\,\mathrm{m}$) using 24 microphones (*Sennheiser KE4-211-2*). The instrument was placed in the center of a sphere with a radius of $1.56\,\mathrm{m}$. Eight microphones were paced on the elevation plane of 0 degree and 45 degree respectively, one microphone was located at each pole (above and below the instrument) and the remaining six microphones were located at the -45 degree elevation plane.

The following instruments were recorded:
Piccolo-trumpet, trumpet, flugelhorn, trombone, violin, viola, clarinet, oboe and transverse flute.
For each instrument, the following tracks were recorded:

- Single tones covering the complete scale of the instrument at different levels

- A short piece of music with a representative pitch-range

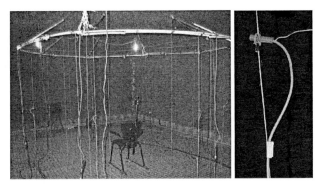

Figure 2.1: Complete measurement setup and a single microphone which is used for the recording.

Additional information and investigations on a very wide range of instruments can be found in Meyer [Mey99]. The main focus here is to generate directivity data being valid for the complete pitch-range of an instrument which can be used for the synthesis of virtual sources.

2.1.2 Analysis

To outline different aspects and their limitations, only two instruments will be discussed as an example for directional radiating instruments such as a trumpet, and rather nondirectional instruments such as a violin. Furthermore, all recordings were done for different levels to estimate the variability of the directivity data with regard to the absolute level of the instrument. The level was increased in steps of approximately 5 dB to cover the whole dynamic range of the specific instrument.

Single Tones

For each single tone a piece of 16,384 samples (371 ms) with approximately constant amplitude is selected from the recorded track ensuring the exclusion of the onset and the decay. This has to be done exact at the same sample position for all directions to ensure a correct time alignment of the channels. After the transformation into the frequency domain, the frequency resolution is 2.7 Hz. Figure 2.2 shows the spectrum of the standard pitch with a level of 90 dB at 2 m. All frequency plots contain the

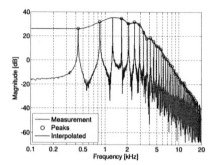

Figure 2.2: Example of generation the continuous frequency spectrum of a trumpet.

fundamental frequency (in this case 440 Hz) and the harmonics with different characteristics dependent on the instrument, the level, and the direction. To generate continuous directivity data, all frequency plots are interpolated using the peaks as sampling points. Finally, the extracted continuous frequency responses of each direction are divided by the frequency response of the reference direction. The result is the normalized directivity.

Interval

As mentioned above, a short interval of the recorded tone is taken for generating each single tone directivity. In this context, tone means the fundamental frequency and all harmonics. However, the tone is not exactly the same (dependent on the player and the instrument) during the complete time interval, even if the on-set and the decay are not considered for the analysis.

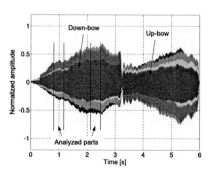

Figure 2.3: Recorded single tone ($fis' = 370$ Hz) of a violin.

To evaluate the influence of this variation, several parts (16,384 samples, i.e. 371 ms length) of the signal (see Figure 2.3) are analyzed and compared. Figure 2.4 depicts the differences for one tone of the trumpet (a) and the violin (b) for one single direction of 90 degree azimuth and 0 degree elevation. The bold curves illustrate the averaging of all different areas of analysis described by the dotted lines. It can be seen that the curves of the trumpet fit almost up to 5 kHz, the violin shows good conformity up to 3 kHz.

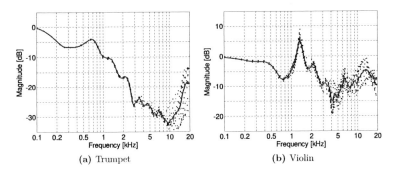

(a) Trumpet   (b) Violin

Figure 2.4: Influence of the selected and analyzed part of a single tone.

Pitch

Up to now, only the directivity of single tones has been analyzed separately. However, the directivity being valid for the complete pitch-range of a specific instrument is required, and thus, the directivity information of all possible tones have to be averaged. Accordingly, it is important to know whether the played tone of the instrument will have any effect on the directivity.

Figure 2.5(a) shows the comparison of the normalized frequency spectra of two different tones played on a trumpet. The solid curve represents the tone $f' = 350$ Hz, the dotted curve the tone $g' = 392$ Hz. The distance in pitch of just 42 Hz does not exactly lead to the same directivity, but the two curves still have a significant similarity. An even stronger influence can be noticed in the case of the violin (Figure 2.5(b)). The solid curve represents the tone $a' = 440$ Hz, and the dotted curve the tone $b' = 466$ Hz. Here, a difference in pitch of only 26 Hz leads to a reversion of the directivity characteristic between 1kHz and 4kHz. This means that a pitch-dependent

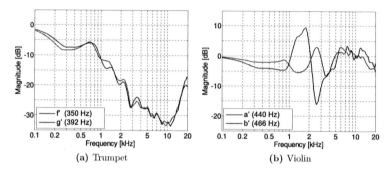

Figure 2.5: Influence of the pitch on the directivity. Direction: 90 degree azimuth and
0 degree elevation.

directivity would be required for an absolute correct representation.

Figure 2.6 shows the set of curves being related to 32 different tones each for a
trumpet 2.6(a) and a violin 2.6(b). The range of tones covers the tones $e = 165$ Hz
to $b'' = 932$ Hz. For the trumpet it is possible to generate a general directivity being
valid for all tones. This is similar for most of the instruments of the same "family"
such as e.g. a piccolo-trumpet, a flugelhorn, a trombone, etc..

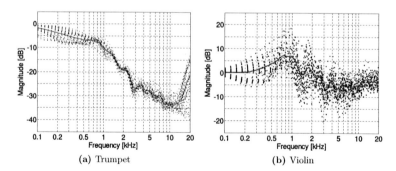

Figure 2.6: Averaging (solid) of 32 different tones (dotted). Direction: 90 degree azimuth
and 0 degree elevation.

The set of curves related to the 32 tones ($g = 196$ Hz to $d''' = 1,175$ Hz) of a violin
shows a distinctive deviation from the average curve.

Analysis of a phrase

A different method to generate a directivity is not to use an averaging over the scale of single tones but, to use a short piece of music with a representative pitch-range. For this reason, several parts of a recorded song were analyzed. These parts with a length of 11.9 s were transformed into the frequency domain and smoothed by using a 1/3 octave window.

Comparing different time intervals of the recorded phrase of a trumpet for the generation of the directivity data shows that the influence is rather low. The plots in Figure 2.7(a) show that all areas produce the same frequency spectrum up to 8 kHz. Except for slight variations of the peaks and dips, all curves fit for the whole frequency range. The differences are almost less than 2 dB. Figure 2.8(a) shows the comparison of the directivity generated with the two different methods. As expected, the directivity curves have nearly the same shape.

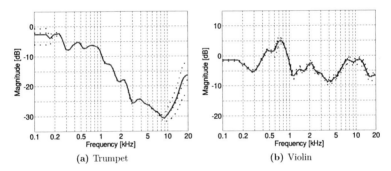

(a) Trumpet   (b) Violin

Figure 2.7: Differences of the directivity according to the analyzed phrase. Direction: 90 degree azimuth and 0 degree elevation.

In contrast to the distinctive deviation of the single tone directivity curves of a violin, the variation of the directivity regarding different time intervals taken for the analysis is much smaller (see Figure 2.7(b)). For this reason, the analysis of a phrase is an effective solution for generating generally valid directivity data, in particular for rather nondirectional instruments such as a violin. It should be taken into account that the pitch-range has to be representative and has to cover all tones which can be played by the specific instrument. Figure 2.8(b) shows the comparison of the directivity generated according to the two different methods. The differences concerning the violin are much higher compared to the trumpet, and the location of the local maxima and minima are more distinctive. In the case of generating the general directivity by averaging the single tone spectra, the high deviation causes a loss of accuracy.

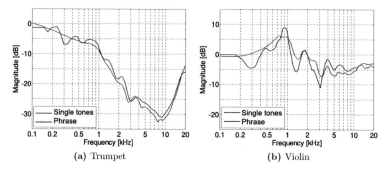

(a) Trumpet   (b) Violin

Figure 2.8: Differences between analysis of a phrase and the averaging of single tones.

An alternative solution proposed by *Rindel* [ROC04, ORC+02] is to use multi-track recordings of each instrument which contain several directions. With this method, signal and directivity are not split and a variation of the spatial radiation dependent on pitch or level are covered by the recorded material. Drawback of this method is the additional effort due to the extended signal processing which rises linearly with the number of channels representing one single instrument. If the number of channels is reduced the spatial resolution is very low. It depends on the individual simulation and the problem which has to be solved to decide which method fits better to the given task. Fortunately, the two methods are not contrary to each other and can be combined in one implementation. This makes it possible to use both dependent on which the simulation is focusing on.

2.1.3   Spatial Interpolation

The generated directivity contains the normalized frequency responses of the 24 recorded directions. However, for a dynamic synthesis process a directivity of a higher spatial resolution is required in order to avoid discontinuities during the movement of the source or the listener. For this reason, the 24 values at each frequency have to be interpolated to a higher spatial resolution. In this case, the resolution is set to

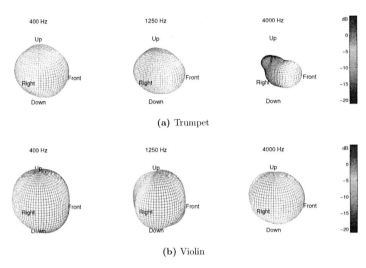

(a) Trumpet

(b) Violin

Figure 2.9: Directivity balloons of a trumpet and a violin at different frequencies. The spatial resolution is interpolated to 5degree. Directivity plots of all measured instruments can be found in the Appendix.

5 degree for both the azimuth and the elevation (see Figure 2.9). Two methods are used for the interpolation. On the one hand, the natural cubic spline interpolation and on the other hand, the piecewise cubic Hermite spline interpolation.

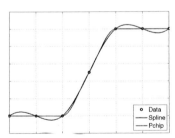

Figure 2.10: Comparison of the two interpolation methods.

The second derivative of the interpolated curve using the natural cubic spline interpolation is continuous for the complete interval. Due to this, it is possible that the interpolated data exceed a local maximum or drop below a local minimum. For the directivity, only the magnitude of the frequency spectrum is taken into account, which therefore only contains only positive values. Tests showed that the natural cubic spline interpolation may result in negative values for the interpolated steps at some frequencies. This occurs especially with instruments with a strong directional radiation such as a trumpet.

In contrast to this method, the slope to the next knot is taken into account by the piecewise cubic Hermite spline interpolation. This leads to the effect that a local minimum remains a local minimum or respectively a local maximum after the interpolation. The second derivative is not necessarily continuous at these points.

2.2   Near-field

Up to now, all examinations regarding the synthesis of sources were related to the assumption, that the source radiates ideally spherical waves similar to a point source but with a directional-dependent amplitude. This is only a theoretical model where the description is only valid for the far-field of the source. There are also other types of sources describing other ideal types of sound radiation. A line source of infinite length has an ideally cylindrical radiation, whereas a planar source produces planar waves. Real sources are three-dimensional sources which can be described - as an approximation - as one of the theoretical and ideal types of sources or a superposition of them in some distance ranges. An array of loudspeakers (line-array) can be described as a line source in between a certain distance range, which is also frequency-dependent. In the far-field the line-array has again a rather spherical radiation, whereas in the near-field of each single loudspeaker the radiated waves are rather planar. Natural instruments are also three-dimensional sources with a very complex radiation of sound. An exact description of the radiation is often difficile and the simulation a time-consuming procedure. Nevertheless, the near-field behavior can be approximated by simple models and still enhances the naturalness and plausibility of the complete simulation.

As an example, a line source with a length of 0.75 m is discussed. The separation between far-field with a spherical radiation and the near-field can be approximated by Equation (2.1). Inside the near-field the level decreases with $1/\sqrt{r}$, in the far-field the level decreases according to the spherical wave attenuation $(1/r)$. The same equation can be used for a piston (planar source) of a certain dimension. In this case l^2 has to be substituted by the surface. The level inside the near-field is approximately constant and distance independent, neglecting the small dips caused by interfering waves from different areas of the piston.

$$r_f \approx \frac{l^2}{\lambda} \quad ; \quad l = \text{Length of the line source} \quad ; \quad \lambda = \text{Wavelength} \qquad (2.1)$$

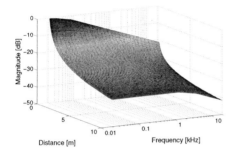

Figure 2.11: Level of a line source as a function of distance and frequency.

Figure 2.11 shows the level of the described line source plotted as a function of the distance and the frequency. It can be seen that for frequencies higher than 100 Hz, the near-field expands to more than 20 cm which is taken here as the closest distance for the simulation. At 20 kHz the range of the near-field is larger than 10 m.

This behavior has a significant influence on the sound of a source in the near-field and can be taken into account by applying a simple frequency-dependent filtering. The cut-off frequency and the frequency dependent attenuation can be calculated by using Equation (2.1) at the specific distance to the source. For a detailed view, the three-dimensional plot in Figure 2.11 is separated into two two-dimensional plots. Figure 2.12 shows the magnitude of the frequency response of the filters at several distances related to that the attenuation in dependence of the distance.

The near-field characteristics are being described by Figure 2.13 and 2.14 in the same manner, but for a planar source with a surface of 0.04 m². It can be seen that the dimension of the near-field is much smaller and has only an influence at very close distances and high frequencies.

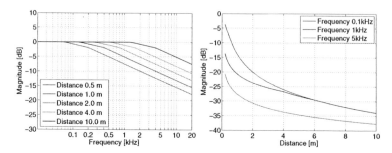

Figure 2.12: Filter frequency response as a function of distance (left) and the attenuation of a single frequency as a function of distance (right) calculated for the line source described in Figure 2.11.

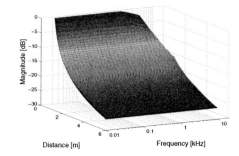

Figure 2.13: Level of a planar source $(S = 0.04\ \mathrm{m}^2)$ as a function of distance and frequency.

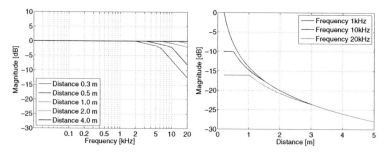

Figure 2.14: Filter frequency response as a function of distance (left) and the attenuation of a single frequency as a function of distance (right) calculated for the planar source.

Chapter 3

Binaural Synthesis

The word "binaural" illustrates that both ears are involved in the perception of a sound event. The different temporal and frequency-dependent cues are interpreted by a complex processing in the brain to reassemble a three-dimensional image of the complete sound field. The processing itself is not discussed further in this thesis but the physical phenomena being responsible for the occurrence of these cues. A special focus is put on the context of dynamic aspects regarding position and orientation of listener and source.

The temporal difference of the signals at the ears is the most important cue for the spatial perception of sound. Furthermore, the sound is diffracted, reflected and shadowed by the head and the torso which also has an influence on the binaural transfer function. In contrast to diffraction, which influences the sound field in the low frequency range, the shading of the head has an influence at high frequencies. All these characteristics are direction-dependent and enable humans to assign a direction to sound events. The ear canal, as the last step of the transmission to the ear drums, brings about a rise of the transfer function at approximately 3 kHz, dependent on the ear canal's individual length. This influence is direction-independent and will not be taken into account for the further discussion of the binaural synthesis. The transfer function to the ears mentioned above is called HRTF, when a representation in the frequency domain is used (Figure 3.1(b)). In the time domain, the equivalent is the Head-Related Impulse Response (HRIR) (Figure 3.1(a)). The time offset between the signals reaching the ears is called Interaural Time Difference (ITD) and can easily be estimated in the time domain plot. By transforming the HRIR into the frequency domain, the Interaural Level Difference (ILD) will be revealed.

In most cases described here, the artificial head developed at the Institute of

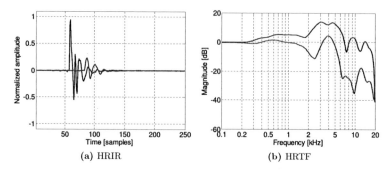

Figure 3.1: Time and frequency domain representation of the transfer functions to the ears for a sound incidence of 70 degree.

Technical Acoustics (ITA) is used. The microphones are placed at the entrance of each ear canal for all measurements, since nearly all direction-dependent filtering is applied to the signal at this point. Tests carried out by [HM91] and [KS93] showed that the suppression of the direction-independent but individually different cues of a binaural signal has a positive effect on the general usability to a wide range of subjects. This point, the entrance of the ear canal, is also the reference point for all further descriptions of binaural measurements, recordings, or reproduction. When measuring the HRTFs of human individuals the same position is used, and furthermore, the ear canal is blocked to obtain approximately the same conditions. A more detailed description of the physics of binaural hearing can be found in [Møl92] and [Bla97].

Figure 3.2: Specified frame of reference for head movements and rotations. According to the application standard in the field of computer graphics the default view direction is along the negative z-axis.

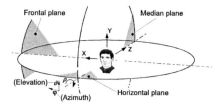

To specify the position and orientation of the head or a source a coordinate system is defined (see Figure 3.2) which is also used in the developed algorithm. In the context of all further descriptions a binaural signal is defined as a two-channel signal containing the left and the right contribution measured at the position above depicted without the influence of the ear canal. For any simulation using the binaural approach it is necessary to divide the binaural signal into the mono audio signal defined by the source and the transfer function to the ears. This separation makes it possible to generate a synthesized binaural signal with any audio signal and an appropriate transfer function to the head (HRTF) which defines the direction of the sound incidence. The procedure of convolving a mono audio signal with an HRTF in order to obtain a synthetic binaural signal is called binaural synthesis. The binaural synthesis transforms a sound source without any position information into a virtual source related to the listener's head. Related to the listener's head means that if the listener moves his head the source will follow this movement. To realize a virtual source which is fixedly related to the room coordinate system a dynamic binaural synthesis is needed. The dynamic system has to choose an appropriate HRTF based on the relative position of the virtual source to the actual position and orientation of the listener's head. This also means that all HRTFs have to be present in the system. It is also possible to realize many different sources and to create a complex three-dimensional acoustical scenario. The spatial resolution which is necessary to realize a smooth changeover while the head or the source is moving is an important topic, which will be discussed in Section 3.3.

3.1 HRTF Measurement

All HRTFs were measured in the semi-anechoic chamber (interior dimensions 6 m $\times$ 11 m $\times$ 5 m; concrete floor) at the Institute of Technical Acoustics using an automated setup. The measurements with the ITA artificial head were carried out with the build-in microphones *Schoeps, CCM 2H*. Measurements of individuals were made using *Sennheiser KE4-211-2* microphones located at the entrance of the blocked ear canals. Individual HRTFs will be used for some evaluations later on. The device under test was mounted (seated) on a turntable, with the ears approximately 2 m above floor level. The measurement signal, a sweep of 16384 samples, was radiated via a *Visaton FR-8-R* full range cone driver (diameter 8 cm) in an enclosed box (volume 0.25 m^3) mounted on a lightweight pendant at a distance of 2 m from the center of the head (see Figure 3.3). Sound azimuth was varied by rotating the turntable. For each azimuth, measurements from different sound elevations were obtained by changing the vertical position of the pendant in successive order, while the source distance was kept constant from the head. All positioning devices were servo-driven and controlled by software. The system measurement, needed for compensation (removal) of the loudspeaker response, was measured at the center position of the head using a 1/2inch free-field microphone (Type 4190, *Brüel & Kjær*).

Figure 3.3:
Automated   HRTF
measurement of the
ITA artificial head
in the semi-anechoic
chamber.

3.1.1 Post Processing

The first-order reflection from the concrete floor in the semi-anechoic chamber had a delay of 6 ms (260 samples at 44.1 kHz sample rate) to the direct signal in the horizontal plane. To eliminate the reflections, a window was applied to all measurements, HRTFs and loudspeaker responses for every elevation angle. After transformation to the frequency domain, the loudspeaker response is eliminated by dividing every HRTF by the loudspeaker measurement at the specific elevation. In spite of the angle-independent sensitivity of the reference microphone (spherical characteristic) the loudspeaker response is measured for every single elevation step to ensure that all HRTFs have absolutely the same time alignment. The lightweight pendant carrying the loudspeaker is leaned toward the device under test at higher elevation angles. Even if the deflection is very low, up to 2 cm at 90 degree, it causes a displacement of the reference point in the head. Having transformed the HRTFs back to the time domain, the time offset according to the distance between loudspeaker and reference point (reference microphone) is added to all resulting HRIRs.

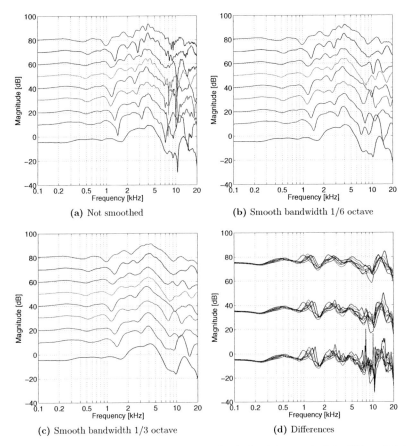

(a) Not smoothed

(b) Smooth bandwidth 1/6 octave

(c) Smooth bandwidth 1/3 octave

(d) Differences

Figure 3.4: Measured HRTFs of different individual subjects and of the ITA artificial head (bold line) at 0 degree azimuth and elevation. Plot (a) without, (b) with applied smooth of 1/6 octave, and (c) with applied smooth of 1/3 octave. Plot (d) shows the differences between the individual HRTFs and the ITA HRTF for the three different cases of smoothing. Curves 3.4(a)–(c) are shifted by 10 dB, groups of curves in (d) are shifted by 40 dB.

Figure 3.4 shows the HRTFs measured of eight individuals and the ITA artificial head. The fine structure of each magnitude spectrum contains many very small notches mainly at higher frequencies. Furthermore, the individual differences are very high with regard to the location and characteristic of the notches. The aim of a binaural synthesis system is to generate spatially distributed virtual sources of a high quality on the one hand, and on the other hand being applicable to a wide range of users without measuring individual HRTFs. The notches can be reduced by a moving average filter, which in this case is a rectangular window of a frequency-dependent variable window size. The window size is fixed according to the aurally accurate logarithmic frequency representation (e.g. 1/6 octave, 1/3 octave), but variable with regard to the underlying linear signal representation (linear FFT spectrum).

The mathematical description of smoothing using a moving averaging with variable window size is depicted in Equation (3.1). It should be noticed that only the magnitude spectrum is smoothed, while the phase information remains unaffected. This is important for obtaining the time alignment ITD between the contralateral and ipsilateral ear. Smoothing decreases the individuality of an HRTF by maintaining the global shape. For this reason, all HRTFs are smoothed with a bandwidth of 1/6 octave to reduce these notches.

$$|H_s(f)| = \frac{1}{f_0 - f_1} \int_{f_0}^{f_1} |H(f)|\, df \quad ; \quad \varphi_{H_s} = \varphi_H \tag{3.1}$$

with

$$f_0 = \frac{f}{\sqrt{2^B}} \quad ; \quad f_1 = f \cdot \sqrt{2^B} \quad ; \quad B = \text{Bandwidth, e.g. } 1/3 \text{ octave}$$

3.2   Near-field HRTFs

An advantage of the binaural synthesis is the ability of near-to-head source imaging. In contrast to panning systems where the virtual sources are always located on or behind the line spanned by the speakers, the realization of a source at any distance to the head can be accomplished with the binaural synthesis by using an appropriate HRTF.

The ability of humans to recognize a source position near to the head is based on distance-dependent differences of the HRTFs. Mainly the interaural characteristics are changed by varying the distance of the source to the head. The distance-dependent characteristics of an HRTF are, as all interaural components, also dependent on the angle of sound incidence.

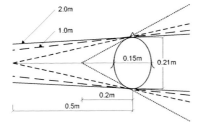

Distance	Angle
0. 2 m	29.8 deg.
0. 3 m	20.3 deg.
0. 4 m	15.3 deg.
0. 5 m	12.3 deg.
0.75 m	8.3 deg.
1. 0 m	6.1 deg.
2. 0 m	3.0 deg.

Figure 3.5: Head shadow of sources at different distances to the head. The dimensions of the head are related to the ITA artificial head.

Figure 3.5 shows the shadow of the head which is increasing for sources at closer distances to the head. This influences mainly the frequency range where the dimension of the head is larger than the wavelength. But also the distance between the point where the direct path of the source to the head is tangential and the contralateral ear is slightly longer which reduces the amount of sound energy at this ear caused by diffraction. The angle range limited by the tangents to the head for the two-dimensional projection used here for simplification is labeled with $2\varphi(r)$. Due to the dependency of r, a possibly not ideal spherical radiation of the measurement loudspeaker has to be taken into account. In this case, it is not possible to compensate the influence of the speaker. This would only be possible for a point to point relation which means that one discrete angle can be defined for the orientation of the source not an angle range (as here $2\varphi(r)$). The only possibility of achieving absolutely correct and comparable results is to use a point source with an ideal spherical radiation. The problem is that most measurement loudspeakers optimized on a spherical radiation suffer when producing a sufficient sound pressure level at the entire frequency range and thus the measured HRTFs have a poor Sound to Noise Ratio (SNR). It should be mentioned here that the generation of HRTFs with an adequate SNR for auralization purposes is much more important than slight influences due to the directivity of the measurement source. Furthermore, all HRTFs should be measured with the same

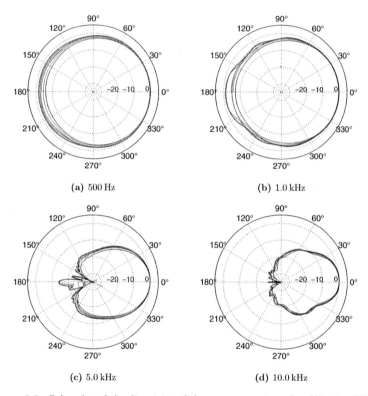

Figure 3.6: Polar plot of the directivity of the measurement speaker (*Visaton FR-8-R*, Ø 8 cm, $V_{box} = 0.25\,\text{m}^3$). All plots are normalized to 0 degree (0 dB).

loudspeaker. Based on these considerations the loudspeaker described in Section 3.1 is also used for the near-field measurements.

To give a more detailed description of the influence of the chosen speaker, Figure 3.6 shows the directivity at frequencies of 500 Hz, 1 kHz, 5 kHz, and 10 kHz at distances of 0.2m, 0.3m, 0.4m, 0.5m, 0.75m, 1.0m, and 2.0m. The frequency response at the different angles are normalized to the frequency response at 0 degree (on axis) and measurement distance, respectively. It is remarkable that the directivity does not depend very much on the measurement distance, at least in the frontal region which is of special interest. Using the approximation (see Equation (3.2)) for the far-field distance of a piston defined by radiation according to the spherical wave attenuation, it can be shown that r_f is about 0.2 m at a frequency of 13.5 kHz.

$$r_f \approx \frac{S}{\lambda} \quad ; \quad S = \text{Surface of the piston} \tag{3.2}$$

At distances closer than r_f (on axis) the sound pressure level remains nearly constant, interrupted by small dips due to the interference of waves radiated by different regions of the membrane. Beyond r_f the sound pressure level decreases according to the spherical wave attenuation. However, it can be seen, that the level is 5 dB lower at 5 kHz compared to the level on axis. For a more detailed description Figure 3.7 shows the frequency response at different distances and the angle $\varphi(r)$ related to the specific distance.

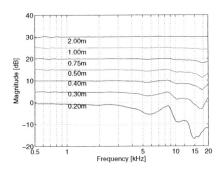

Figure 3.7: Frequency responses of the measurement speaker for each distance at the specific angle α. All curves are shifted by 5 dB.

Figure 3.8 shows exemplarily HRTFs measured at distances of 0.2 m, 0.5 m, and 2.0 m. It can be seen that at closer distances and at higher frequencies the ILD is much higher. But even at lower frequencies of about 100 Hz, the level difference between the ipsilateral and contralateral ear at 0.2 m is much higher compared to the HRTF at 2.0 m. At 100 Hz the wavelength is about 3.4 m, and thus these frequencies are not much affected by the shadowing effect. Assuming a point source or at least a source with a partially spherical radiation, the relative distance difference between the contralateral and the ipsilateral ear is important due to the effect of the spherical wave attenuation. This is the substantial factor causing a higher ILD also at low frequencies. Figure 3.9 shows this dependency calculated (see Equation (3.3)) for a sphere approximating the head.

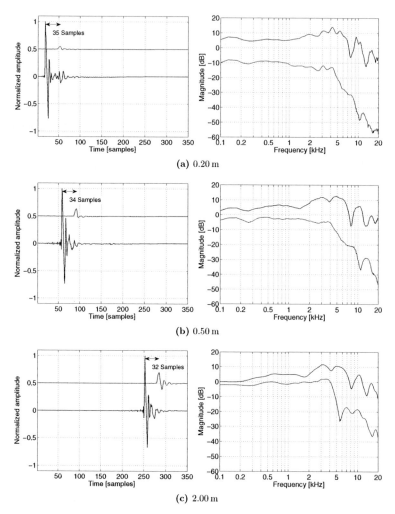

Figure 3.8: Near-field HRIRs (left side) and HRTFs (right side) measured at distances of 0.2 m, 0.5 m, and 2 m. The azimuth angle is 90 degree in the horizontal plane. All plots are normalized to the time domain maximum of the ipsilateral ear and smoothed by 1/6 octave in the frequency domain. The curves of the contralateral ear are shifted by 0.5 V in the time domain plots.

The ITD remains almost unchanged as can be seen in the time domain plots in Figure 3.8. The maximum ITD varies between 32 samples (726 μs) at a distance of 2.0 m and 35 samples (794 μs) at the closest distance of 0.2 m . However, this minor deviation (3 samples ≈ 70 μs) has to be kept in mind for a possible interpolation between distances. This will be discussed later in Section 3.3 *Dynamic Aspects*.

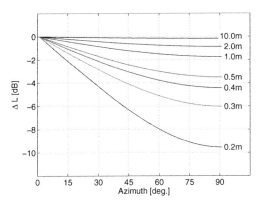

Figure 3.9: Spherical wave attenuation $\Delta L(\varphi, r)$ related to different source distances to the center of the head plotted for angles between 0 degree (frontal direction) and 90 degree (lateral direction).

$$\Delta L(\varphi, r) = 20 log \frac{r - d/2 sin(\varphi)}{r + d/2 sin(\varphi)} \qquad \begin{array}{l} \varphi = \text{Azimuth (0 deg. = frontal direction)} \\ d = \text{Ø head} \\ r = \text{Distance to the head (center)} \end{array} \qquad (3.3)$$

Among others, Brungart and Rabinowitz [BR99] made detailed investigations on the auditory localization of near-to-head sources. These investigations focused on distances between 0.12 m to 1.0 m. To differentiate the regions of near and far distances, the terms "proximal region" and "distal region" were introduced. The same nomenclature will be used here for the further distance-related discussion on HRTFs. Brungart and Rabinowitz defined the boundary between the distal and proximal region at a distance of 1.0 m to the center of the head.

As mentioned above, it is possible to realize every distance of a virtual source, in particular, one close to the listener's head, by choosing the appropriate HRTF which provides all cues being necessary for a proper identification. However, the primary aim on which the focus is put on in this thesis is the realization of a dynamic system where the sources or the head are intended to change their position and orientation. In this sense, the determination of the distances at which the differences between the distal and proximal region can be perceived during a filter change is a substantial issue.

Furthermore, for practical reasons of implementation, it is neither feasible nor necessary to store HRTFs for every possible distance, and if a perceptual perimeter of the proximal region can be found, it will be possible to reduce the amount of HRTFs that have to be present in the system. Inside the proximal region special HRTFs have to be used together with an appropriate interpolation at distances between two measured distances. Beyond the proximal region, the sound pressure level decreases in the same range at both ears and, distances can be synthesized by using HRTFs measured at one distance with a simple level adaption. Waves emitted from a source located in the distal region have approximately spherical characteristics, the ILD is nearly constant, and both ITD and ILD are distance-independent. The level decreases according to the spherical wave attenuation for the specific distance. So, the entire dependency on the distance can be modified simply during the runtime of the program.

3.2.1   Listening Test

To evaluate the perceptual perimeter of the proximal region, a simple listening test was performed in the semi-anechoic chamber, examining the ranges where different near-field HRTFs have to be applied [Phe02]. The listeners were asked to compare signals from simulated HRTFs with those from correspondingly measured HRTFs on two criteria, namely the perceived location of the source and any coloration of the signals. In this listening test only the horizontal plane was taken into account. A variation of the elevation angle was omitted as the major differences related to a variation of source distance can be found at lateral angles where the maximum distance between both ears occurs. At any elevation angle outside the horizontal plane, the interaural distance decreases and also the distance-dependent differences [BDR99].

The HRTFs for the listening test were measured in the semi-anechoic chamber at distances between 0.2m and 1.0m with a spacing of 0.1m. Between 1.0m and 2.0m, the spacing was chosen to 0.25 m. The azimuth angle was modified in steps of 15 degree between the frontal (0 degree) and rear (180 degree) direction. A preliminary test showed, as expected, that differences become more significant in the lateral area. An approximately similar behavior was observed in front and beyond the frontal plane. To reduce the number of trials, the tested angles were limited to 0 degree, 45 degree,

75 degree, and 90 degree. The simulated HRTFs were prepared from far-field HRTFs (measured at a distance of 2 m) with a level correction applied to both channels.

From the comments of all 9 listeners, for each distance and azimuth, a coloration would earn 5 points, while a change of the perceived location earned 10 points. No perceived difference earned 0 points. The total points were then added up. Therefore, the higher the total score, the greater the perceived audible difference between the original and the simulated HRTFs for the specific distance. Figure 3.10 shows the result. The nearer boundary line defines where huge audible differences and a change of the perceived location of the sound were detected. The boundary line being further away from the head is the boundary where any audible difference such as slight coloration was detected. As expected, differences between the simulated distance and the original distance were reported mainly for stimuli in the lateral region. Already at a distance of 1.5 m at 90 degree, slight differences were audible. In this context this fact is most important for the further design of the systems database. The first point where a detection of differences occurs defines the boundary line between the proximal and distal region. Within the proximal region, a near-field representation is necessary by using special HRTFs. The distal distance of 1.5 m used here differs from the value (1.0 m) which is to be found in literature. The measured differences between sources at 1.5 m and 1.0 m are not very significant, which is presumably the reason for defining the upper boundary of the proximal region at 1.0 m. But to avoid any audible artifacts at the transition of the two regions, the perception-based boundary of 1.5 m is used here as the basis for the HRTF database design.

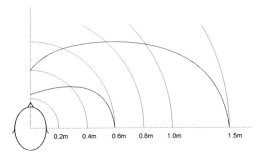

Figure 3.10: Limits of noticeable differences between near-field and far-field HRTFs.

For the synthesis database, the HRTFs of the ITA head were measured at distances of 0.2 m, 0.3 m, 0.4 m, 0.5 m, 0.75 m, 1.0 m, 1.5 m, and 2.0 m [Len07]. The spatial resolution is chosen to 1 degree for the azimuth angle and 5 degree for the elevation angle. The closer spacing between measurements at distances near to the head has to be applied to minimize any comb-filter effects and irregularities concerning the perceived distance. Consequently, the generated HRTF database covers all requirements for a near-to-head source imaging system. It should be noted that the system uses the HRTFs of the full sphere because the ITA head has asymmetrical pinnae and head geometry. Non-symmetrical pinnae have positive effects on the externalization of the generated virtual sources [BT05].

3.3   Dynamic Aspects

The importance of head movements for improved localization is well known and reported by many authors (e.g. [TR67, WK99]). As stated above, the position of a virtual source generated by using binaural synthesis implies that the source moves with the listener. For realizing a room-related virtual source, the HRTF has to be adapted when the listener turns or moves his head. Thus, besides the generation of room-related sources, the main advantage of a dynamic synthesis is an almost complete elimination of front-back confusion as it often appears when using static binaural synthesis with non-individualized HRTFs (see for example [WAKW93]).

Figure 3.11 shows the interaural time differences in relation to the listener's orientation. In this example, the ITD is almost equal regardless whether the sound is reproduced at position (1) or (2). Although the frequency-dependent ILD is still different for the two source positions, this is often inadequate as an unique cue. For that reason, the signal could be perceived by the user as if coming from a non-existent mirror source due to the congruent differences in the ITD. If the listener moves his head, the ITD will increase when the listener's frontal direction turns away from source (1) and will decrease when the frontal direction turns toward source (2). Due to this fact, a source is well defined at its position due to the ancillary information, i.e. the relative movement of the listener.

Despite the advantage of the dynamic synthesis in view of better localization results, the technical drawback is the filter change being required in order to achieve a room related virtual source. Whenever a source is moving or the head moves relative

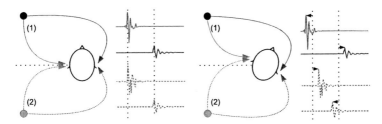

Figure 3.11: Variance of the interaural time difference depending on the relative head orientation of the user.

to a fixed source, the transfer functions from the source to the ears change [LAVK06]. In the case of binaural synthesis, this means that the HRTFs used for the filtering have to be changed, but every filter change also causes artifacts in the resulting audio signal.

There are several important factors that have to be comprised to ensure a smooth inaudible changeover of HRTFs:

- Spatial resolution (angle increment between two HRTFs)

- Fading method

- Fading time

- Update time

Unfortunately, these factors are partially predicated among each other, and thus, for a better understanding, the technical prerequisites will be described separately in this chapter. In Chapter 6, where the complete VR system is introduced, the chosen parameters will be discussed according to the constrains which are imposed by the system's architecture, the computational requirements, and the usage of the complete system. There are two fundamental aspects concerning the problem of a smooth filter changeover. Either the differences of the filters are small and the artifacts are not audible while switching between HRTFs which is attributed to the spatial resolution of the HRTF database and the update time of the entire system or an adequate fading method has to be used to suppress these artifacts.

3.3.1   Spatial Resolution

The maximum angle increment between two HRTFs without producing any audible artifacts can only be investigated by applying a listening experiment. As a test signal a standard white noise signal is used as input audio signal to be filtered. Standard white noise is a broadband signal with randomized phase response, and it is highly suitable as a test signal since colorations and clicking are perceived immediately [Bla05]. The test should either substantiate or disprove the assumption that a simple switch, without any form of cross-fading, between two filters of highly similar characteristics does not result in any audible artifacts.

Figure 3.12 shows the left ear HRIRs from 0 degree to 359 degree, measured in the horizontal plane in azimuth steps of 1 degree (a) and the ITD as a function of azimuth for different elevation angles (b). It is evident that the HRIRs from within the same vicinity in space have more or less the same shape, differing mainly in time delay. The listening test was carried out with different angle increments between 1 degree and 5 degree. Increments above 5 degree were not investigated due to possible jumps in the spatial localization [Bla97, PS90].

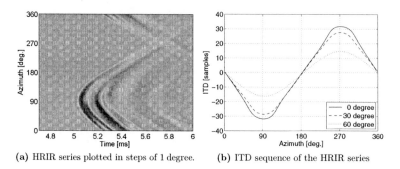

(a) HRIR series plotted in steps of 1 degree.   (b) ITD sequence of the HRIR series

Figure 3.12: Measured HRIRs of the horizontal plane and the associated ITD sequence.

In a first step, the test signal is filtered with each of the 360 left and right-ear HRIRs to generate a rotating source through all 360 degree with an increment of 1 degree. The results are as expected; there were no audible artifacts at all during the transition from one filter to the next. The test is repeated, with applied azimuth increments of more than 1 degree. Switching between filters in azimuth steps of 2 degree and more give rise to audible click noises, increasing in audibility as the azimuth steps are increased. The most critical region of switching is the region for lateral sound incidences. In particular, the artifacts are perceived at the contralateral ear which is also congruent with the findings of [HP04]. A consequence thereof is that without any cross-fading or interpolation technique, a spatial resolution of 1 degree is required.

In a PC-based dynamic synthesis system, the storage of an HRTF database in a spatial resolution of 1 degree is not the problematic factor but rather that it has to be ensured that the filter change itself is not carried out with increments of larger steps as 1 degree. Unfortunately, the maximum angle threshold is not defined by the system, rather it depends on the position and orientation data delivered by the head-tracking device. Tracing devices, such as they are used in the application described in Section 6, *Interactive VR-System*, work with update rates of about 60 Hz. This would result in a maximum speed of a relative angle change of 60 deg./s too. If the listener turns his head with a speed of more than 60 deg./s, a filter change with an increment of 1 degree cannot be guaranteed. As a consequence, additional mechanisms as cross-fading and interpolation have to be considered and implemented in the system for an artifact-free filter changeover to override the limitation to a restricted movement velocity. This will be discussed in the following sections. However, during a movement with low velocity a simple filter switch is sufficient, as shown by the listening experiment, which saves computational power.

3.3.2 Fading

The aim of the previous section was to find an upper boundary for the differences between two HRTFs without any perceptible artifacts. If for any reason an interchange from one HRTF to the next with an increment of more than 1 degree is required an adequate technique is needed to change from the actual filter to the next. A fading from one filter to the next is the simplest method and requires less computational resources compared to interpolation. Nevertheless, it doubles the amount of operations during the fading process due to the parallel processing of two filters for one output. Fading represents the summation of both filtered audio signals each with a weighting factor as a function of time. The result processed with the actual filter decreases in the defined fading time slot, and the result processed with the next filter is increased accordingly. The filter characteristic itself is not interpolated.

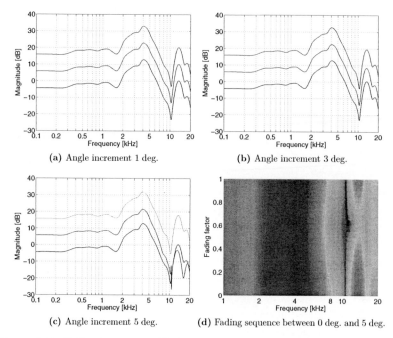

(a) Angle increment 1 deg.

(b) Angle increment 3 deg.

(c) Angle increment 5 deg.

(d) Fading sequence between 0 deg. and 5 deg.

Figure 3.13: Fading between two HRIRs plotted with different angle increments. The results are plotted in the frequency domain.

Figure 3.13 depicts the start and target of the fading process which are the measured HRTFs and the result at the time when both HRTFs have the same weighting factor in the middle of the fading process. The uniform changeover plotted in Figure 3.13 (a) is desirable but not remarkable, when using this small increment of 1 degree and is only given here as an anchor for comparison proposes. An angle increment of 3 degree or 5 degree represents fading conditions which are more usual in a dynamic system.

The result supposes that a simple fading is still sufficient for an angle threshold of 3 degree (see Figure 3.13(b)). However, Figure 3.13(c) and (d) show that the summation of the two HRIRs results in a decrease of the level and in interfering dips especially at higher frequencies and in the middle of the fading process. Thresholds of more than 3 degree may result in audible artifacts and, thus, an interpolation is required. These assumptions were also confirmed in listening experiments.

3.3.3 Interpolation

As stated above, angle increments of several degrees produce dips in the frequency domain, which can be perceived as a sound coloration during the fading process. This is mainly caused by the phase misalignment of the two HRTFs. Basically, the problem can be solved in the time or the frequency domain. Both methods have different advantages.

Time Domain

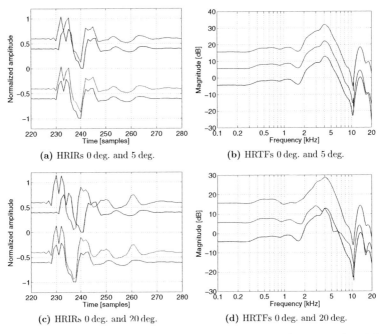

(a) HRIRs 0 deg. and 5 deg. (b) HRTFs 0 deg. and 5 deg.

(c) HRIRs 0 deg. and 20 deg. (d) HRTFs 0 deg. and 20 deg.

Figure 3.14: HRIR and HRTF of 5 deg. and 20 deg. respectively compared to 0 deg.. Curves in (a) and (c) are shifted by ±0.5 (pair) and ±0.1. Curves in (b) and (d) are shifted by 10 dB.

As an example, the HRIR of 0 degree is compared to the HRIR of 5 degree and 20 degree respectively. Figure 3.14(a) shows the HRIRs of the contralateral ear plotted native (upper curves) and after the time delay compensation (lower curves). The curves with a congruent time alignment have very similar shapes and amplitudes. In contrast to Section 3.3.2 the fading using the time-aligned HRIRs gives very good results, as can be seen in the frequency domain in Figure 3.14(b). Basically, interpolation allows changing the filter characteristics gradually to avoid problems using the simple fading technique up to higher angle increments (see Figure 3.15).

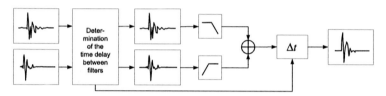

Figure 3.15: Principle of interpolation in the time domain.

By knowing the time-delay, the two filters can be aligned with each other in the time domain. Now it is possible to fade the preceding filter gradually to the succeeding filter. Synchronously the resulting filter is shifted to its correct position. The same issue is described in Figure 3.14(a) and (b) but for an angle increment of 20 degree. In this case, the main peak in the time domain plot is still similar, but the superposition shows an irregular frequency response, and as a consequence the limitation of this method.

A fundamental problem of this method is a correct determination of the time delay and its gradual manipulation. A simple registration of the maximum of each filter does not guarantee a correct alignment among the filters, as shown in Figure 3.16. It is possible that the second peak of the one HRIR is the maximum, whereas the other HRIR has its maximum at the first peak. On the opposite, the cross-correlation of the two HRTFs is a well suited and robust method to determine the time delay correctly (Figure 3.16(b)).

Due to the range of the time delay of one or at most very few samples focused on here, a gradual shift has to be processed also in sub-sample steps. Besides a realization by using an all-pass filter, a phase manipulation can be processed in the frequency domain. Here, the phase can be easily adjusted by modifying the relation of the real

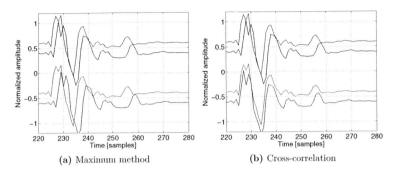

(a) Maximum method   (b) Cross-correlation

Figure 3.16: Time alignment estimated with the maximum method versus cross-correlation.

and imaginary part of each sample. With these preliminary considerations it is an obvious conclusion to process the complete interpolation in the frequency domain.

Frequency Domain

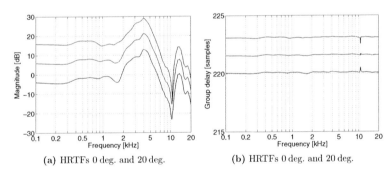

(a) HRTFs 0 deg. and 20 deg.   (b) HRTFs 0 deg. and 20 deg.

Figure 3.17: Interpolation in the frequency domain using HRTF of 0 degree and 20 degree. Curves in (a) are shifted by 10 dB.

Figure 3.17(a) depicts the result for an interpolation in the frequency domain by using two HRTFs with an angle increment of 20 degree. The interpolated HRTF (curve in the middle) is calculated by using only the magnitude of both, the start and target HRTF with the same weighting factor. The result is remarkably better than when using the time domain interpolation shown in Figure 3.14(c) in the previous section.

To ensure a correct time delay of the resulting filter, an additional step is required after the interpolation of the magnitude. The time delay of a filter is represented in the frequency domain as the group delay, the first derivative of the phase response. The group delay of a filter response such as an HRIR is a function of frequency and indicates the time delay of each frequency component. Mathematically, the group delay of a filter with the phase response of $\varphi(f)$ is defined as:

$$t_g = -\frac{1}{2\pi} \cdot \frac{d\varphi(f)}{df} \tag{3.4}$$

The definition of the group delay implies that the interpolation has to be done by modifying the unwrapped phase response of the filter. In contrast to the simple shift of the impulse response in the time domain, which represents an uniform shifting of the group delay, the interpolation in the frequency domain enables a frequency-dependent shifting of the group delay. It is, however, not allowed that the frequency response is being modified by by the group delay shift. This means that the magnitude of the complex values must be kept constant, whereas only the arguments are modified accordingly (see Figure 3.18). Furthermore, by using a spherical spline interpolation separately for magnitude and phase instead of weighting, the results are even better for large angle increments.

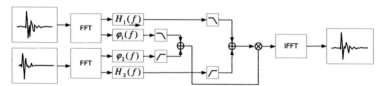

Figure 3.18: Principle of interpolation in the frequency domain.

It should be mentioned here, that regardless of which interpolation is used, every step between the HRTFs is only a linear superposition of both. An HRTF measured in comparison to an interpolated HRTF in the same angle, may consist of dips and peaks at slightly different frequencies which cannot be represented by the interpolation. The variation of the filters' fine structure is lost in between the interpolated angle region. Particularly, HRTFs at lateral angles show a higher variation than HRTFs of the frontal direction. For this reason, interpolation is to be considered as a method to reduce audible artifacts, when large angle increments are inevitable, but not to reduce the spatial resolution of the HRTF database.

3.3.4   Distance Interpolation

As discussed in Section 3.2 *Near-field HRTFs*, it is advisable to use special HRTFs
within the proximal region. These HRTFs already contain the cues needed to
realize a convincing near-field perception of virtual sources. However, some further
consideration will be necessary if a dynamic solution is desired, meaning in this
case, the variation of distance to the virtual source within the proximal region. A
variation of the distance induces an interpolation between two HRTFs at different
measured distances to realize an arbitrary virtual distance of the source. In contrast
to the proximal region, the ILD caused by lateral sound incidence from a source in
the far-field is dominated by the head shadowing effect not by the distance difference
between the ears. The sound pressure level at low frequencies is rather the same.
Applying a simple level correction according to the spherical wave attenuation to
both channels of the HRTF is a sufficient approximation. As an example, when a
source is shifted from 2 m to 3 m, the angle to the head is $\varphi = 90$ degree (lateral
sound incidence). The HRTF for the synthesis is measured at a distance of 2 m. The
error that occurs when applying the simple level correction according to the spherical
wave attenuation is only ≈ 0.14 dB. As implied by Figure 3.9 in Section 3.2, the level
difference between both ears for lateral sound incidence becomes important with
a decreasing distance to the head and can not be neglected. Due to the distance
dependency of the level difference, the interpolation between distances is more
complex in this case.

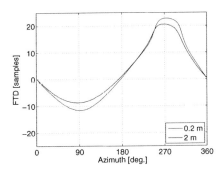

Figure 3.19: Time difference of the ear
signals related to the frontal direction
of sound incidence as a function of the
head angle for source distances of of
0.2 m and 2 m.

The time delay or distance difference respectively, of each ear in relation to the frontal direction as a function of azimuth and elevation is an useful value for considerations about distance interpolation. Figure 3.19 shows this time difference of the ITA artificial head and in addition the time difference calculated by using the spherical head model at the left ear for different elevation angles. The time difference related to the frontal direction will be referred to as Frontal Time Delay (FTD) in all further descriptions. The difference between the FTD value at both ears is identical with the ITD for each angle of sound incidence. In contrast to the ITD which is rotationally symmetric to the point ($\varphi = 180$ degree, $t_{ITD} = 0$ s), the FTD shows a strong asymmetry which reflects the indirection of the sound waves to the contralateral ear. Furthermore, the FTD calculated by using the spherical model is a good approximation to the measured values.

$$r_{contralateral,near} = \varphi \cdot \cos(\vartheta) \cdot r_H + \sqrt{r^2 - r_H} \qquad (3.5)$$

$$r_{ipsilateral,near} = \begin{cases} \varphi \cdot \cos(\vartheta) \cdot r_H + \sqrt{r^2 - r_H} & \varphi \geq \dfrac{\arcsin(r_H/r)}{\cos(\vartheta)} \\[3ex] \sqrt{(r - r_H \sin(\varphi))^2 - (r_H \cos(\varphi))^2} & \varphi < \dfrac{\arcsin(r_H/r)}{\cos(\vartheta)} \end{cases} \qquad (3.6)$$

$$r_{contralateral,far} = r + r_H \cdot \sin(\varphi) \cdot \cos(\vartheta) \qquad (3.7)$$

$$r_{ipsilateral,far} = r - r_H \cdot \sin(\varphi) \cdot \cos(\vartheta) \qquad (3.8)$$

Figure 3.21 shows the level at the ears as a function of distance for different angles of sound incidence. The diameter of the head is chosen to 18 cm. The level is calculated by using two different spherical models given with Equation (3.5) to (3.8). Equation (3.5) and (3.6) represent the exact solution for any distance with respect to the assumption of a spherical head. This is delineated by the solid curves in

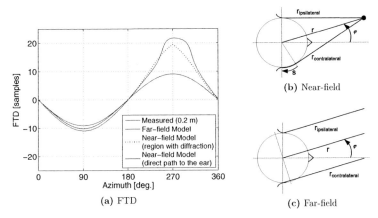

Figure 3.20: Frontal time delay (FTD) calculated using the two different head models ((b) and (c)) in comparison to the measured FTD for a source distance of 0.2 m.

Figure 3.21. Equation (3.7) and (3.8) constitutes an approximation disregarding the indirection of the sound waves due to diffraction (dotted curves). Points labeled with ∗ and ○ represent the real level of the measured HRTFs at 100 Hz. It can be seen, that the exact model fits well with the measured values at all distances, whereas the approximation gives acceptable results at distances of 2 m and beyond at least at the contralateral ear. More information about spherical models can be found in [MPO+00, SHLV99, AAD01].

The synthesis database contains HRTFs of the ITA head measured at distances of 0.2 m, 0.3 m, 0.4 m, 0.5 m, 0.75 m, 1.0 m, 1.5 m, and 2.0 m, All distances between the measured values have to be interpolated by using the Equations (3.9) and (3.10). Due to the distance dependency of the FTD (see Figure 3.19) the interpolation has to be done separately for magnitude and phase in the same way as described in Section 3.3.3 to avoid any audible artifacts. The interpolation of the magnitude can be divided into two subparts. In a first step, both HRTFs will be scaled separately for the left and the right channel according to the corresponding distance calculated by using the spherical model (Equation (3.5), (3.6)).

With the individual scaling factor, the global level of the two involved HRTFs can be adjusted. This can be interpreted as operating point for the weighing factor which controls the degree of influence of each HRTF according to the distance. This way, a

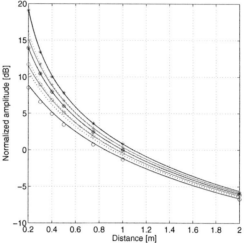

Figure 3.21: Level at the ears as a function of distance for different angles of sound incidence (center curve: $\varphi = 0$ and outer curve: 90 degree). The level is calculated using a sphere model of the head. The diameter of the head was chosen to 18 cm. Points marked with $*$ and $\circ$ represent the real level of the measured HRTFs at 100 Hz.

smooth changeover of all characteristics to be considered is ensured, i.e. level, phase, and magnitude.

$$\left| HRTF_{L/R} \right| = \left| HRTF_{1,L/R} \right| \cdot a_{Scale,1} \cdot a_{Weighting} \tag{3.9}$$
$$+ \left| HRTF_{2,L/R} \right| \cdot a_{Scale,1} \cdot \left(1 - a_{Weighting} \right)$$

$$\varphi_{L/R} = \varphi_{1,L/R} \cdot a_{Weighting} + \varphi_{2,L/R} \cdot \left(1 - a_{Weighting} \right) \tag{3.10}$$

with:

$$a_{Scale,1} = \frac{r_{1,L/R}}{r_{L/R}} \quad ; \quad a_{Scale,2} = \frac{r_{2,L/R}}{r_{L/R}} \quad ; \quad a_{Weighting} = \frac{r_{2,L/R} - r_c}{r_{2,L/R} - r_{1,L/R}} \cdot \frac{r_{1,L/R}}{r_c}$$

$r_{L/R}$ = Actual distance to the left/right ear

$r_{1,L/R}$ = Distance to the left/right ear to the next lower measured HRTF

$r_{2,L/R}$ = Distance to the left/right ear to the next upper measured HRTF

Chapter 4

Crosstalk Cancellation

All information required for a correct spatial impression is provided by a binaural signal. If it is possible to reproduce the binaural signal exactly at the eardrums, a listener should percept no differences between the reproduction of a recorded or simulated sound field and a real sound field. Headphones and loudspeaker are both feasible to reproduce a binaural signal.

The major advantage of using headphones is the system-induced channel separation between the left and right ear. Headphones, however, also have several disadvantages which go beyond technical difficulties. First of all, a proper binaural reproduction requires an exact equalization of the headphones to eliminate the transfer function of the headphones to the point of the ear canal at which the binaural signal is valid for [Vor00]. This refers to the location of the microphones during the recording of the signal or the measurement of the HRTFs. It is difficult, due to the high variability, to obtain one filter characteristic which is valid generally for all users [KC00]. The externalization of virtual sound images without additional signal processing in headphone reproduction is often a problematic issue, too [KC05]. Nevertheless, in spite of all possible corrections being made accurately, listening over headphones is often deemed not equivalent to free-field listening [RF95].

Apart from the possibility to reproduce a binaural signal over headphones, the loudspeaker reproduction is an effective an appropriate method of signal reproduction without using any wearable hardware. The main problem of loudspeaker reproduction, however, is the crosstalk between the channels, which spoils the three-dimensional cues of the binaural signal [Møl89]. This chapter describes the technique of reproducing binaural signal over loudspeakers but saving the spatial cues by suppressing or at least reducing the interfering crosstalk.

First of all a description of the concept of crosstalk cancellation in a static implementation is given to provide an overview of the basic mathematical background. Furthermore, the limitation (*sweet spot* = valid area) of the static solution is evaluated to obtain a base line for further considerations. Section 4.2 *Dynamic Solution* describes the extension of the static solution to an dynamic adaptive solution to override the limitation of the *sweet spot*.

4.1 Static Solution

As suggested above, the essential prerequisite for a correct binaural presentation is that the right channel of the signal is audible only in the right ear and the left channel only in the left ear. This procedure is called crosstalk cancellation and was introduced by ATAL and SCHRÖDER [AS63]. With an adequate filtering this crosstalk effects can be compensated and a sufficient separation both channels can be achieved [Bau63]. Abbreviation used to describe the principle of crosstalk cancellation are used as follows:

X_L, X_R: Input signals
Y'_L, Y'_R: Speaker input signals
Y_L, Y_R: Speaker signals
Z_L, Z_R: Ear signals
H_{LL}:   Transfer function from left speaker to left ear
H_{LR}:   Transfer function from left speaker to right ear
H_{RL}:   Transfer function from right speaker to left ear
H_{RR}:   Transfer function from right speaker to right ear
L: Transfer function of the left speaker
R: Transfer function of the right speaker

Figure 4.1 depicts the iterative structure of the cancellation process. The impulse (1) which is provided to be audible at the left ear only, is repeated (2, 3) over both loudspeakers according to the temporal conditions and the level reduction of the transmission to achieve a destructive superposition with the crosstalk (and again the crosstalk of the compensation steps) at the ears. The four transfer functions describing

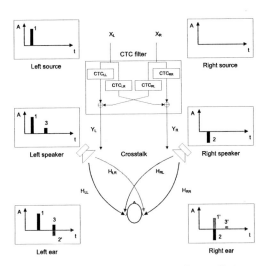

Figure 4.1: The principle of crosstalk cancellation.

the transmission from each loudspeaker to each ear are denoted by H_{LL}, H_{LR}, H_{RL}, and H_{RR}. Theoretically, the iteration has to be accomplished to an infinite number of steps, but due to the decreasing level of the higher order steps, the iteration can be stopped after approximately 5 to 7 steps. Detailed information on the number of steps required for a sufficient compensation can be found in [Sch93].

A correct binaural reproduction is achieved if the following conditions are complied. The complete transfer function from the left input of the compensation network to the left ear (reference point is the entrance of the ear canal) including the transfer function H_{LL} has a flat frequency response. The same applies for the right transfer path, accordingly. The crosstalk indicated by H_{LR} and H_{RL} has to be canceled by the system.

For a mathematical consideration of this problem, the closed solution of [Møl89] is more suitable than the iterative solution, which is more demonstrative to discuss the principle. Both solutions can be transferred into each other, which will be shown later on.

The ear signals Z_L and Z_R can be described as:

$$Z_L = Y_L \cdot (L \cdot H_{LL}) + Y_R \cdot (R \cdot H_{RL}) \qquad\qquad = X_L \qquad (4.1)$$

$$Z_R = Y_R \cdot (R \cdot H_{RR}) + Y_L \cdot (L \cdot H_{LR}) \qquad\qquad = X_R \qquad (4.2)$$

For a correct binaural reproduction the condition $Z_L = X_L$ and $Z_R = X_R$ must be true. The two unknowns Y_L and Y_R are brought on the left side by solving Equation (4.1) and (4.2):

$$Y_L = \frac{1}{L} \cdot \left(\underbrace{\left(\frac{H_{RR}}{H_{LL} \cdot H_{RR} - H_{LR} \cdot H_{RL}} \right)}_{CTC_{LL}} \cdot X_L - \underbrace{\left(\frac{H_{RL}}{H_{LL} \cdot H_{RR} - H_{LR} \cdot H_{RL}} \right)}_{CTC_{RL}} \cdot X_R \right)$$
$$(4.3)$$

$$Y_R = \frac{1}{R} \cdot \left(\underbrace{\left(\frac{H_{LL}}{H_{LL} \cdot H_{RR} - H_{LR} \cdot H_{RL}} \right)}_{CTC_{LR}} \cdot X_R - \underbrace{\left(\frac{H_{LR}}{H_{LL} \cdot H_{RR} - H_{LR} \cdot H_{RL}} \right)}_{CTC_{RR}} \cdot X_L \right)$$
$$(4.4)$$

Y_L and Y_L represent the signals distributed by the loudspeaker and contain each both input signals convolved with a filter in a way that the crosstalk paths H_{LR} and H_{RL} will be compensated at the ears. The four CTC blocks can be described as followed:

$$CTC_{LL} = \frac{H_{RR}}{H_{LL} \cdot H_{RR} - H_{LR} \cdot H_{RL}} \qquad (4.5)$$

$$CTC_{LR} = -\frac{H_{LR}}{H_{LL} \cdot H_{RR} - H_{LR} \cdot H_{RL}} \qquad (4.6)$$

$$CTC_{RL} = -\frac{H_{RL}}{H_{LL} \cdot H_{RR} - H_{LR} \cdot H_{RL}} \qquad (4.7)$$

$$CTC_{RR} = \frac{H_{LL}}{H_{LL} \cdot H_{RR} - H_{LR} \cdot H_{RL}} \qquad (4.8)$$

$$(4.9)$$

Figure 4.2 shows the block diagram of the crosstalk cancellation network.

The closed solution is accurate and in theory the perfect solution for the complete CTC, but assumes a filter with an infinite step response. This is only possible, if recursive filters are used, which are not mandatory stable in all cases. Furthermore it is not

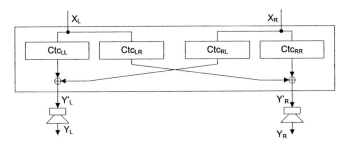

Figure 4.2: Block diagram of the CTC filter set.

possible to process a recursive filter from measured impulse responses (here H_{LL}, H_{LR}, H_{RL}, and H_{RR}) with an absolute correct characteristic of phase and magnitude as it is essential in this application. For this reason FIR-Filter with a finite length, which is defined by the impulse response of the transmission path, are used. To convolve the audio signal with the filters calculated by the closed solution requires a windowing of the filter impulse response. In contrast to that, using the iterative method, the length of the filter response can be determined in advance. With every iteration, the step response extends about the interaural time difference [Sch93, NMS92]. Transferring the closed solution into the iterative solution is described in the following section, starting with only one branch of the compensation network. It should be mentioned that the differences between both methods decrease when the length of the FIR filter is increased.

$$
\begin{aligned}
CTC_{LL} &= \frac{H_{RR}}{H_{LL} \cdot H_{RR} - H_{LR} \cdot H_{RL}} \\
&= \frac{\frac{H_{RR}}{H_{LL} \cdot H_{RR}}}{\frac{H_{LL} \cdot H_{RR}}{H_{LL} \cdot H_{RR}} - \frac{H_{LR} \cdot H_{RL}}{H_{LL} \cdot H_{RR}}} \\
&= \frac{1}{H_{LL}} \cdot \frac{1}{1 - \frac{H_{LR} \cdot H_{RL}}{H_{LL} \cdot H_{RR}}}
\end{aligned}
\tag{4.10}
$$

The following notation will be used:

$$
K = \frac{H_{LR} \cdot H_{RL}}{H_{LL} \cdot H_{RR}}
\tag{4.11}
$$

The closed solution can be transferred to an iterative with a geometric series:

$$\frac{1}{1-K} = \sum_{m=0}^{\infty} K^m \qquad \forall \quad |K| < 1 \tag{4.12}$$

The blocks of Figure 4.2 can be described with these equations:

$$CTC_{LL} = \frac{1 + K + K^2 + \dots}{H_{LL}} \tag{4.13}$$

$$CTC_{RR} = \frac{1 + K + K^2 + \dots}{H_{RR}} \tag{4.14}$$

$$CTC_{LR} = -CTC_{LL} \cdot \frac{H_{LR}}{H_{RR}} \tag{4.15}$$

$$CTC_{RL} = -CTC_{RR} \cdot \frac{H_{RL}}{H_{LL}} \tag{4.16}$$

For the measurement of the static CTC system (see Section 5.1) the geometric series was proceeded up to the 2nd power. The CTC blocks (4.13) and (4.14) were built like this:

$$CTC_{LL} = \frac{1 + K + K^2}{H_{LL}} \tag{4.17}$$

$$CTC_{RR} = \frac{1 + K + K^2}{H_{RR}} \tag{4.18}$$

4.1.1 Sweet Spot

The static crosstalk cancellation is defined only for a single point, but the area around this point in which still a sufficient channel separation is being achieved is called the *sweet spot*. The dimension of the *sweet spot* will be analyzed in this section as an determining factor for the design of a dynamic crosstalk cancellation. It is known from several studies before, that the dimensions of the *sweet spot* are not very large [Ama00, RNRT02]. A dynamic working system has to adapt within the *sweet spot* on a changed listener position to ensure that the differences regarding coloration and localization are below a noticeable level. All six degrees of freedom have to be considered to determine what kind of movements is most critical. Furthermore this provides a baseline for the required spatial resolution of the HRTF database.

Setup

A listening test is carried out to determine the dimension of the *sweet spot* defined
by the perceptive parameters *localization* and *coloration*. The filter set of the static
CTC is designed for a specific point which represents the center of the listener's
head. This point is also the center of the *sweet spot* and thus the reference point
for the listening tests [Buc04]. The test person were asked to report changes in the
perceived localization and coloration of the presented stimuli during the specified
movement. The test person indicates whether he or she has detected a change
and then the current head position is logged. The test is carried out in the semi-
anechoic chamber at the ITA. Figure 4.3 shows a schematic arrangement of the setup.

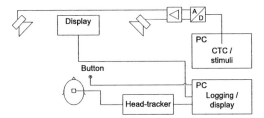

Figure 4.3: Block diagram of the listening test setup.

The following kinds of movements are analyzed:

- *Lateral*: lateral movement along the *x*-axis

- *Azimuth*: rotational movement about the *y*-axis

- *Elevation*: rotational movement about the *x*-axis

- *Roll*: rotational movement about the *z*-axis

Only the reaction regarding a lateral movement is analyzed here, as this is the
most critical movement compared to the vertical movement (along the *y*-axis) or the
movement back and forth (along the *z*-axis). A previous study [Ama00] has shown
that these movements are rather uncritical or that the *sweet spot* is more expanded
in these directions respectively. The test person wears a lightweight helmet with a
head tracking receiver to determine the head position and orientation precisely at

the time-stamp the button is pressed by the test person to report a perceived change
A *Flock Of Birds* is used as tracking device [Asc]. Furthermore the tracking data
is used as a visual feedback for the test person. Deviations are displayed separately
for all degrees of freedom on a monitor placed in the reference view direction. This
helps the test person to move only in the specified direction, and thus, reduces the
undesirable dependence between the movements. Three different types of stimuli were
used to determine the *sweet spot*: a pulsed broadband noise, low-pass filtered noise
pulses, and a continuous broadband noise. The pulsed noise is appropriate to test
changes in the localization whereas the continuous signal is more likely to detect any
colorations depending on the deviation to the center of the *sweet spot*. The virtual
sources generated by using the binaural synthesis (also static) are placed lateral at
±90 degree. The test person decides whether the left or the right stimulus is used
during the test.
The following stimuli are used:

- *Pulse*: three pink noise pulses of 350 ms length separated by 250 ms silence and
  determined by 700 ms silence, played in a loop

- *Pulse, LP*: low-pass filtered (1.5 kHz) pink noise pulses of the same sequence as
  above

- *Continuous*: four second continuous pink noise, determined by one second silence, played in a loop

Results

The box plots and the corresponding statistics depicted in Figure 4.4 show the results
of the listening test separately for each degree of freedom and the sign of the movement. The percentile values indicate the amount of test persons who report changes
at a certain deviation. A 25th percentile at a deviation of 2 cm means that 25% of the
test persons perceived a change in the signal either in terms of coloration or a shift
of the source location. The boxes in the box plots are spread from the 25th percentile
(bottom) to 75th percentile (top).

The variability of the results for lateral movements shown in Figure 4.4a regarding
the type of stimulus or the sign of the movement (back and forth) are rather small.
This means that there is not one specific stimulus or the sign of the movement that

turned out to be as more critical than others. The mean values are between $\approx 1.5\,\mathrm{cm}$ and $\approx 3.0\,\mathrm{cm}$ which substantiates the assumption and findings of other studies that the dimension of *sweet spot* is not very large. The rotational movement about the y-axis (azimuth) is most important for the localization of sound sources and thus, also of special interest for the crosstalk cancellation (Figure 4.4b). The mean values are between $\approx 10\,\mathrm{degree}$ and $\approx 20\,\mathrm{degree}$. The mean values regarding the rotational movement about the x-axis (elevation) start at $\approx 17\mathrm{degree}$ and extend to $\approx 24\mathrm{degree}$, which is much higher than for the azimuth. Also the standard deviation is significantly higher. The value ranges from $\approx 8\mathrm{degree}$ to $\approx 15\mathrm{degree}$. The results for the rotational movement about the z-axis are again in the same range as for the azimuth. This indicates that the crosstalk cancellation is much more sensitive to deviations that cause reverse time structures at the ears. In general, the test subjects tend to detect changes while moving toward the sound source at a lower deviation as when moving away from the source. For all different stimuli, moving toward the source produces a lower mean value as well as a lower standard deviation.

The primary goal of this study was to find expressive dimensions of the *sweet spot*. An interesting finding is that the range of deviation differs depending on the type of stimulus and the sign of the movement. However, it is not that important for the further development of a dynamic crosstalk cancellation system but the fact that the lowest dimension is being found. As a baseline the dimensions of the *sweet spot* are summarized in Table 4.1, which contains the values of the lowest 10th percentile and of the lowest mean value of all different types of stimuli.

Movement	Lowest 10th percentile	Lowest median
Lateral	0.6 cm	1.5 cm
Azimuth	4.4 deg.	10.4 deg.
Elevation	4.7 deg.	17.5 deg.
Roll	1.3 deg.	12.8 deg.

Table 4.1: "Inner" dimension of the *sweet spot*

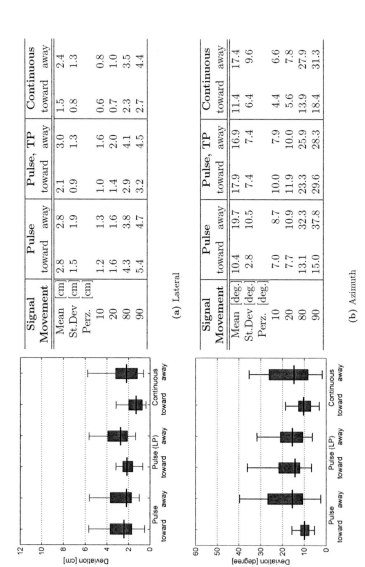

Signal Movement	Pulse		Pulse, TP		Continuous	
	toward	away	toward	away	toward	away
Mean [cm]	2.8	2.8	2.1	3.0	1.5	2.4
St.Dev [cm]	1.5	1.9	0.9	1.3	0.8	1.3
Perz. [cm]						
10	1.2	1.3	1.0	1.6	0.6	0.8
20	1.6	1.6	1.4	2.0	0.7	1.0
80	4.3	3.8	2.9	4.1	2.3	3.5
90	5.4	4.7	3.2	4.5	2.7	4.4

(a) Lateral

Signal Movement	Pulse		Pulse, TP		Continuous	
	toward	away	toward	away	toward	away
Mean [deg.]	10.4	19.7	17.9	16.9	11.4	17.4
St.Dev [deg.]	2.8	10.5	7.4	7.4	6.4	9.6
Perz. [deg.]						
10	7.0	8.7	10.0	7.9	4.4	6.6
20	7.7	10.9	11.9	10.0	5.6	7.8
80	13.1	32.3	23.3	25.9	13.9	27.9
90	15.0	37.8	29.6	28.3	18.4	31.3

(b) Azimuth

Figure 4.4: Box plot of the *sweet spot* for different movements, each for different types of stimuli.

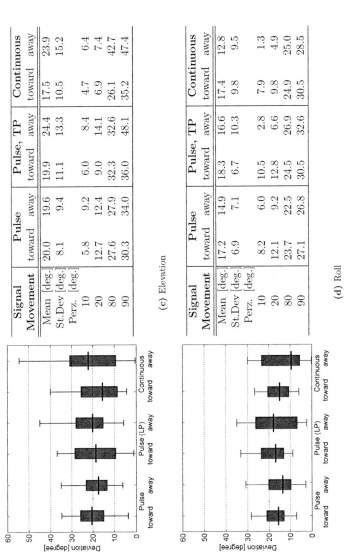

Signal Movement	Pulse toward	away	Pulse, TP toward	away	Continuous toward	away
Mean [deg.]	20.0	19.6	19.9	24.4	17.5	23.9
St.Dev [deg.]	8.1	9.4	11.1	13.3	10.5	15.2
Perz. [deg.]						
10	5.8	9.2	6.0	8.4	4.7	6.4
20	12.7	12.4	9.0	14.1	6.9	7.4
80	27.6	27.9	32.3	32.6	26.1	42.7
90	30.3	34.0	36.0	48.1	35.2	47.4

(c) Elevation

Signal Movement	Pulse toward	away	Pulse, TP toward	away	Continuous toward	away
Mean [deg.]	17.2	14.9	18.3	16.6	17.4	12.8
St.Dev [deg.]	6.9	7.1	6.7	10.3	9.8	9.5
Perz. [deg.]						
10	8.2	6.0	10.5	2.8	7.9	1.3
20	12.1	9.2	12.8	6.6	9.8	4.9
80	23.7	22.5	24.5	26.9	24.9	25.0
90	27.1	26.8	30.5	32.6	30.5	28.5

(d) Roll

Figure 4.4: Box plot of the *sweet spot* for different movements, each for different types of stimuli.

4.2   Dynamic Solution

Using the method for eliminating the crosstalk between the two loudspeakers at the ears of the listener, discussed in the previous section, it is possible to calculate a filter set for every location of the listener by knowing the corresponding transfer functions. Thus, a correct reproduction of a binaural signal is possible at every point. A dynamic operating crosstalk cancellation solution which has the ability to adapt on the listener's position requires as a precondition, that the system can access the listener's current position and orientation. For this reason the listener has to be tracked by a head-tracking device. The technical requirements regarding update-frequency or latency are discussed later on in Section 6 *Interactive VR-System*. In addition to that, the system must provide a valid filter set for "every" point inside the listening area. This filter calculation can be carried out either on-line or off-line. An advantage of the off-line calculation is that very little computational power is required during runtime and thus, the very short update time, which is only limited by the access time of the memory [Len06].

Off-line filter calculation would be possible but is not applicable for a system which should cover the listening area, i.e. of a *CAVE-like* environment. The dimension of the *sweet spot* indicates that the spatial distance between two filters may not be too large. For example: The three-dimensional listening area should be 2 m × 2 m × 1 m with a vertex distance of 2 cm each with rotational components ($\Delta\varphi = \pm 180$ degree, $\Delta\vartheta = \pm 60$ degree, $\Delta\rho = \pm 30$ degree) would cause the total amount of $\approx 10^{12}$ filter sets (four filters each). By using a filter length of 1024 samples, this would cause the total amount of ≈ 10 PByte memory. Apart from the high amount of memory, it is necessary to prepare a database for every environment the system is used for.

The approach presented in this section focuses on a filter calculation during run-time. In this case, only the HRTFs required as input data for the filter calculation have to be stored. GARDNER [Gar97] demonstrated the general applicability in his studies. The database which has to present in the system contains only HRTFs measured with a proper spatial resolution at a certain distance, which is discussed in the next section.

Having acquired all information regarding position and orientation it is possible to recover every transfer function between loudspeaker and the ear of the listener and to calculate the filters on-line, though this solution requires much more computational power. Nevertheless, a dynamic solution is being possible which preserves a valid filter set for the listener's current position.

4.2.1   HRTF Database

The starting point for the filter calculation is the designation of the appropriate HRTFs regarding the current relative position of the listener to the speakers. The listener's position is provided by the head-tracking device, the loudspeaker positions have to be present in the system. It is obvious, that the spatial resolution of the HRTF database does not always match calculated relative angles exactly. In this context it is interesting to know which spatial resolution meets the requirements regarding the dimension of the *sweet spot* and therewith the ability to change the filters without audible artifacts.

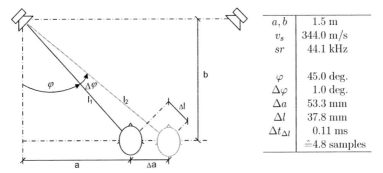

a, b	1.5 m
v_s	344.0 m/s
sr	44.1 kHz
φ	45.0 deg.
$\Delta\varphi$	1.0 deg.
Δa	53.3 mm
Δl	37.8 mm
$\Delta t_{\Delta l}$	0.11 ms
	$\hat{=}$4.8 samples

Figure 4.5: Simple example to evaluate the spatial resolution of the HRTF database required for the CTC filter calculation.

Figure 4.5 depicts a simple example to consider of which spatial resolution the HRTFs have to be measured to comply first the requirements given by the *sweet spot*. The dimensions of the listening area defined by the speaker setup are similar to the ones found in environments the system is considered for, e.g. a *CAVE-like* environment or a *L-Bench*. At least the order of magnitude is the same. The example shows that a lateral movement of $\approx$ 5 cm just causes a change of one degree in the relative orientation of the listener to the loudspeaker. The dimension of the sweet spot discussed in Section 4.1.1 regarding a lateral movement is determined even lower to 1.5 cm, which is corresponding to a change of just 0.28 deg. in the relative orientation. This implies a spatial resolution of 1/4 deg. or higher. An examination of the variations in the HRTFs regarding frequency and time seems to be useful to ensure, that the resolution has to be that fine.

Figure 4.6 depicts the differences in the magnitude of HRTFs between 45 and 46 deg measured in steps of 1/4 deg. All curves are normalized to 45.0 deg. It can be seen that the differences in the frequency domain are very small for increments inbetween 1 deg. This means that the information content regarding the frequency response does not change significantly. Compared to this, the plot in the time domain reveals some differences in the time alignment, which obviously cause the decrease of the channel separation apart from the center of the *sweet spot*. Due to the considerations made above a HRTF database with a resolution of 1 deg for both, azimuth and elevation, is used as a basis for the dynamic crosstalk cancellation.

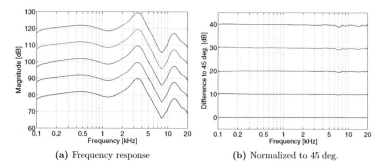

(a) Frequency response   (b) Normalized to 45 deg.

Figure 4.6: Frequency response and differences of HRTFs of 45.0 to 46.0 deg. measured in steps of 1/4 deg.. Only the contralateral ear is plotted.

4.2.2   Filter Calculation

All database queries are rounded to the next integral angle (depicted as $(\varphi, \vartheta)_{round}$ in Equation (4.19)). All HRTFs were measured with the loudspeaker at a fixed distance to the head and thus, contain a fixed time offset beside the angle dependent ITD. This time offset has to be adapted with respect of the listener's current distance to each reproduction loudspeaker.

The following equation depicts the time offset correction:

$$h_{(\varphi,\vartheta)}(t) = h_{(\varphi,\vartheta)_{round}}(t) * h_\Delta(t)$$

$$= \underbrace{h_{(\varphi,\vartheta)_{round}}(t) * h_{\Delta,samples}(t)}_{\hat{=}shift} * h_{\Delta,fragment}(t)$$

$$= h_{(\varphi,\vartheta)_{shifted}}(t) * h_{\Delta,fragment}(t) \tag{4.19}$$

$$\circ\!\!\!\bullet$$

$$H_{(\varphi,\vartheta)}(f) = H_{(\varphi,\vartheta)_{shifted}}(f) \cdot H_{\Delta,fragment}(f) \tag{4.20}$$

After determining the required time correction $h_\Delta(t)$ the integral offset $h_{\Delta,samples}(t)$ (largest integral number of samples) and the fragment offset $h_{\Delta,fragment}(t)$ are split. This offset is positive or negative depending on whether the distance of the listener to the speaker is larger or smaller than the measurement distance. The displacement in time expressed mathematically by the convolution with a shifted dirac pulse can be replaced for the integral part by a simple circular shift of the impulse response. As mentioned above, the fragment part has to be taken into account for a correct cancellation. All filters will be transformed into the frequency domain to complete the calculation. This avoids the time consuming convolution. After the transformation into the frequency domain, the fractal part of the time correction is a simple phase manipulation which is already described in Section 3.3.3 *Interpolation*. All required transfer functions are now prepared for the CTC filter calculation as described in Section 4.1.

4.2.3 Filter Post-Processing

The time alignments of the resulting CTC filters are influenced by the time offsets of the HRTFs used for the calculation. The time offset of a HRTF represents the distance between the head and the loudspeaker. The filter processing uses a filter length of 2048 samples. Inside the filter boundaries the HRTFs can be shifted according to the distance to the loudspeakers. Thus, the chosen filter length affects also the size of the valid listening area. The drawback is the resulting latency due to the offset between the first sample and the main peak which means that the filter has a higher group delay. Besides, a longer filter is computationally more expensive, especially for the continuous convolution of the input signal.

To reduce the latency of the system and also the computational load the complete filter set is being truncated after the calculation to 1024 samples (see Figure 4.7) or 512 samples depending on the desired range of the listening area. The time alignments among the four filters are not modified. The truncation can be interpreted as shifting all loudspeakers virtually towards the listener to the same extend.

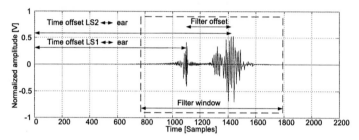

Figure 4.7: Applied window to the calculated CTC filters. The calculation length is set to 2048 samples, the resulting filter length for the convolution is set to 1024 samples.

4.2.4 Stability

Tests have shown that the dynamic CTC works very well within the angle spanned by the loudspeakers using the straightforward method described above. Outside this area some problems occur, such as enormous sound colorations, ringing, and even range overflow. For a better examination and analysis it is helpful to convert Equation (4.5) - (4.8) into a representation that is more related to the iterative structure of the alternating compensation steps shown in Figure 4.1. One term (Equation (4.11)) can be extracted from each of the four terms representing the CTC filters, which can be transformed to a geometric series, where K is used as a substitution for $(H_{LR} \cdot H_{RL}) / (H_{LL} \cdot H_{RR})$. K is associated with the iteration corresponding to the cancellation model. A detailed description can be found in [Sch93].

If for any frequency $H_{LL} \cdot H_{RR}$ is equal to $H_{LR} \cdot H_{RL}$, K becomes 1 and the filter is no longer stable using the straightforward method. Even if the result of $1/(1 - K)$ is not exactly but very close to zero, the resulting filter reaches a high amplitude, which may result in ringing or even a range overflow at the sound output device. In such cases it is possible to use an adaptive filter, such as a Wiener filter, to avoid instability problems. This technique is implemented with many variations in some applications. A detailed description of adaptive filter techniques can be found, for instance, in [KW03] and [MK99]. This study is aiming at an optimized system performance of the complete dynamic CTC. In this context a deeper analysis of the reasons and effects of the instability is worthwhile. Besides several singularities at some frequencies, which can be avoided, it can be seen that the amplitude of the frequency vector K is related to the filter decay in the time domain. As a concrete description, the behavior of a geometric series applies. The geometric series converges faster for small values than for large values. As far as the cancellation model is concerned, this means that the filter decay time is longer for values of K close to 1. From the mathematical point of view this fact does not affect the cancellation performance, but in the context of a real error-afflicted reproduction, the quality of the sound is affected significantly. Under real listening conditions the HRTFs present in the system's database are not identical to the listener's HRTFs. Furthermore there may be slight inaccuracies in the head position and in the determination of the orientation, which might be caused by the tracking device itself or by an incorrect placing of the tracking receiver or the position markers relative to the ears. It should be noted that the point of reference for the head tracking is a single point at the user's head. The transformation of this position to the user's ear is not the same for every user due to the different geometries of human heads. Another mismatch can be produced by the system's latency. If the listener moves fast, the latency of the head tracker and the additional processing time of a new filter set can cause a slight displacement of the canceling signal, that is, the superposition is not completely destructive. All these displacements in the filter's time structure can be minimized but not eliminated under real working conditions in a VR environment.

It is, however, important that there is a connection between the filter decay time and the audibility of these errors. A long decay time of the CTC filters, which can be associated with more iteration steps of the geometric series that have to be considered for a proper cancellation, causes more errors and therefore a smaller channel separation than a filter with a short decay time. Moreover, if more signal energy is needed for the cancellation and the environment is not ideally anechoic, the reflections contain also more energy. This causes audible artifacts and a loss of channel separation.

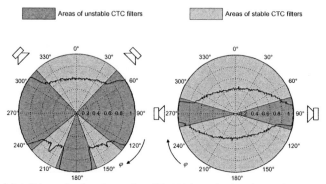

(a) ±45 degree loudspeaker configu- (b) ±90 degree loudspeaker configuration
ration

Figure 4.8: Plots of the highest values of K (below 1) for every head orientation,
$|K_{max}(\varphi)|$. Both loudspeaker and ears are placed 1.7 m above floor.

The values of the frequency vector K depend in general on the orientation of the
head relative to the loudspeakers. The shading of the head causes a level decrease at
the contralateral ear. For head orientations within the angle range spanned by the
loudspeakers, H_{LR} and H_{RL} are always the transfer functions to the contralateral ear
whereas H_{LL} and H_{RR} describe the transmissions to the ipsilateral ear. Outside this
area the association of the contralateral and ipsilateral ear to one loudspeaker changes.
The borderline is defined by the angle at which the head is facing a loudspeaker. Then
H_{RR} becomes a transfer function to the contralateral ear and H_{RL} to an ipsilateral
ear for a ±45 degree loudspeaker setup and a clockwise head turn. This means that
both loudspeakers "see" the same ear as the ipsilateral ear. In this area the crosstalk
is higher, and so is the value of K due to the reversed influence of the head shadow.

In addition to keeping the timbre as constant as possible during movements of
the listener, robustness of the system is a primary goal in this development. Given
the considerations made above, the straightforward method is chosen for the realization of the spatial audio system described in Section 6 *Interactive VR-System*. This
method requires the lowest computational costs and extends the number of loudspeakers from two to four to cover all possible regions with stable filters, without a
need for integrating other stabilizing filter techniques.

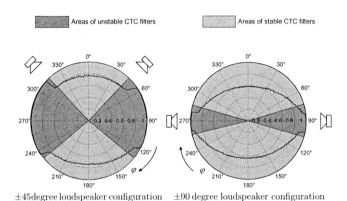

Figure 4.9: Plots of the highest values of K (below 1) for every head orientation, $|K_{max}(\varphi)|$. Loudspeakers are placed 3.0 m, ears 1.7 m above floor.

In order to detect possible singularities, K is calculated for every head direction. Afterwards the algorithm searches the highest value in each filter and stores this value for the current head position. The highest value of K indicates also whether the resulting CTC filter will have a shorter or a longer decay time.

The polar plots in Figures 4.8 and 4.9 show the results of the search for critical K values over a complete head rotation for different loudspeaker configurations and heights. The display range of the plots is limited to 1. The areas according to head orientations that do not produce a value above 1 are valid areas. Figure 4.8(a) illustrates that a stable CTC is possible in a range of approximately ±40degree around the frontal direction using the ±45 degree loudspeaker configuration. In Figure 4.8(b) (±90 degree configuration), the valid area is about ±75 degree around the frontal direction.

It is not necessary to place the loudspeakers in the plane of the listener's ears. They can also be mounted in an elevated plane. Choosing the adapted HRTFs from the database to calculate the compensation filters produces the same sound pressure at the eardrums of the listener. This fact is a strong argument for using this technique in a *CAVE-like* environment, where placing the loudspeakers is the main problem. In most cases it is only possible to place them in the upper corners. The areas of stability produced by the two different loudspeaker configurations mounted at a height of 3 m are shown in Figure 4.9.

4.2.5   Dual Crosstalk Cancellation

A four loudspeaker environment turns out to be favorable as it provides a complete 360 degree rotation with a stable filter set at every possible viewpoint of the listener. This makes it possible to combine the ±45 degree configuration (sectors I, III, V, VII in Figure 4.10) and the ±90 degree configuration (sectors II, IV, VI, VIII in Figure 4.10) in the same setup. Using four loudspeakers eight combinations of a normal stereo CTC system are possible. However, the validity areas of the two-channel CTC systems overlap. Depending on the current viewpoint, the system automatically chooses the valid configuration. Thus the current binaural audio signal is filtered with the correct CTC filter for the listener's present position. Measurements and listening impressions showed that the cancellation achieves good results within specific areas, but switching between areas is still audible as a "click". Comparing CTC filters one step before and one step after switching them reveals some differences, in particular at high frequencies. As mentioned before, small inaccuracies regarding the position of the loudspeakers and the determination of the head position by the head tracker can cause a mismatch in the time alignment in a way that the filter changeover is not sufficiently consistent at high frequencies.

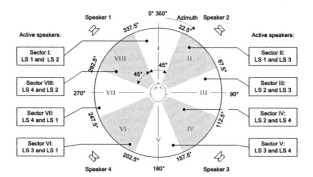

Figure 4.10: Sectors of the four speaker CTC System. Grey areas ±90degree configuration; white areas ±45 degree configuration.

Sector Fading

A smoother changeover from one sector to the next is needed to reduce the interfering clicks. The first solution was a short fading within one output block of the sound output device (256 samples, i.e. 5.8 ms at a 44.1 kHz sample rate) as it is also used for the changeover between the areas. Tests showed that the audible clicks could be eliminated completely, but more than half of the test subjects reported a short jump of the virtual source toward the closest activated loudspeaker. After a moment or after turning the head only within one sector, the virtual source was again stable at its intended position. The signal still contains uncompensated artifacts, not related to the virtual sources, which can be heard due to the precedence effect when the direction of incidence changes abruptly. This happens when the loudspeaker setup is changed within a very brief time span. Therefore a method was developed to distribute the fading process to a wider angular dimension. In the following this technique is referred to as dual crosstalk cancellation. The four-loudspeaker setup introduced above is described by Equation (4.21) and (4.22).

$$Z_L = Y_1 \cdot H_{1L} + Y_2 \cdot H_{2L} + Y_3 \cdot H_{3L} + Y_4 \cdot H_{4L} \qquad (4.21)$$
$$Z_L = Y_1 \cdot H_{1R} + Y_2 \cdot H_{2R} + Y_3 \cdot H_{3R} + Y_4 \cdot H_{4R} \qquad (4.22)$$

This set contains four unknowns but only two equations, which makes a closed solution impossible [BC96]. It is only possible to give a numerical solution, which is an approximation to an exact solution. Furthermore calculating the filters in that way is CPU-time intensive.

In the VR environments focused on the dynamic CTC system has to react very quickly to the listener's movements, and it must be robust in any situation. Due to the fact that the CTC filter structure is a linear system, a linear superposition of two classical two loudspeaker systems is possible. Based on the switching method described above, an alternate use of the two (± 45 degree and ± 90 degree) loudspeaker configurations, namely a superposition of both within the overlapping area, can be developed. The following nomenclature will be used to describe the procedure. We define an active area and a destination area. In each case only one loudspeaker configuration has to be active to provide a sufficient and stable CTC.

Loudspeakers in the active area are labeled A and B and those in the destination area are labeled A and C. While fading from sector I to sector II, for example, loudspeakers 1 and 2 (A, B) are active. After the fading loudspeakers 1 and 3 (A, C) are active. Thus, a generally applicable system of equations can be established,

$$Z_L = Y_A \cdot H_{AL} + Y_B \cdot H_{BL} + Y_C \cdot H_{CL} \tag{4.23}$$

$$Z_R = Y_A \cdot H_{AR} + Y_B \cdot H_{BR} + Y_C \cdot H_{CR} \tag{4.24}$$

with

$H_{source\,drain}$ = transfer function from source to drain.

The set of Equation (4.23) and (4.24) contains again more unknowns than equations, making a closed solution impossible. Taking into account the boundary conditions before and after fading, this set of equations can be split into two independent cancellation problems, which will be superimposed. A complete channel separation, and therefore a perfect CTC, is true if $Z_L = X_L$ and $Z_R = X_R$. In this case the signals at the listener's ears and the binaural input signal are identical. Because of the system's linearity the input signal and one loudspeaker signal each can be split into two parts:

$$X_L = X_L^{AB} + X_L^{AC} \quad ; \quad X_R = X_R^{AB} + X_R^{AC} \quad ; \quad Y_A = Y_A^{AB} + Y_A^{AC} \tag{4.25}$$

Here X_L^{AB} is the part of the left input signal reproduced by loudspeakers A and B. Y_A^{AB} is the components of the signal at loudspeaker A related to the corresponding signal at loudspeaker B. This becomes clear when the terms in Equation (4.25) are inserted into Equations (4.23) and (4.24):

$$Z_L = Y_A^{AB} \cdot H_{AL} + Y_B \cdot H_{BL} + Y_A^{AC} \cdot H_{AL} + Y_C \cdot H_{CL} \tag{4.26}$$

$$Z_R = Y_A^{AB} \cdot H_{AR} + Y_B \cdot H_{BR} + Y_A^{AC} \cdot H_{AR} + Y_C \cdot H_{CR} \tag{4.27}$$

It is now possible to split each equation into two parts and solve them separately. The resulting systems represent two "classic" CTC structures. Both systems contain two unknowns and two equations and so a closed solution is possible.

System AB:

$$Z_L = Y_A^{AB} \cdot H_{AL} + Y_B \cdot H_{BL} \tag{4.28}$$

$$Z_R = Y_A^{AB} \cdot H_{AR} + Y_B \cdot H_{BR} \tag{4.29}$$

$$Y_A^{AB} = \underbrace{\left(\frac{H_{BR}}{H_{AL} \cdot H_{BR} - H_{AR} \cdot H_{BL}}\right)}_{CTC_{LL}^{AB}} \cdot X_L - \underbrace{\left(\frac{H_{BL}}{H_{AL} \cdot H_{BR} - H_{AR} \cdot H_{BL}}\right)}_{CTC_{RL}^{AB}} \cdot X_R \tag{4.30}$$

$$Y_B^{AB} = \underbrace{\left(\frac{H_{AL}}{H_{AL} \cdot H_{BR} - H_{AR} \cdot H_{BL}}\right)}_{CTC_{LR}^{AB}} \cdot X_R - \underbrace{\left(\frac{H_{AR}}{H_{AL} \cdot H_{BR} - H_{AR} \cdot H_{BL}}\right)}_{CTC_{RR}^{AB}} \cdot X_L \tag{4.31}$$

System AC:

$$Z_L = Y_A^{AC} \cdot H_{AL} + Y_C \cdot H_{CL} \tag{4.32}$$

$$Z_R = Y_A^{AC} \cdot H_{AR} + Y_C \cdot H_{CR} \tag{4.33}$$

$$Y_A^{AC} = \underbrace{\left(\frac{H_{CR}}{H_{AL} \cdot H_{CR} - H_{AR} \cdot H_{CL}}\right)}_{CTC_{LL}^{AC}} \cdot X_L - \underbrace{\left(\frac{H_{CL}}{H_{AL} \cdot H_{CR} - H_{AR} \cdot H_{CL}}\right)}_{CTC_{RL}^{AC}} \cdot X_R \tag{4.34}$$

$$Y_C^{AC} = \underbrace{\left(\frac{H_{AL}}{H_{AL} \cdot H_{CR} - H_{AR} \cdot H_{CL}}\right)}_{CTC_{LR}^{AC}} \cdot X_R - \underbrace{\left(\frac{H_{AR}}{H_{AL} \cdot H_{CR} - H_{AR} \cdot H_{CL}}\right)}_{CTC_{RR}^{AC}} \cdot X_L \tag{4.35}$$

A factor depending on the actual orientation of the listener's head will be defined to fade from one sector to the next. Each sector limit will be replaced by a small fading area. Again the active sector before entering the fading area is only driven by system AB. At the beginning of the fading the actual head angle φ is equal to φ_0; at the end it equals φ_1.

Conditions at the start of fading $(\varphi = \varphi_0)$,

$$X_L = X_L^{AB} \quad ; \quad X_R = X_R^{AB} \quad ; \quad Y_A = Y_A^{AB} \tag{4.36}$$

Conditions at the end of fading $(\varphi = \varphi_1)$,

$$X_L = X_L^{AC} \quad ; \quad X_R = X_R^{AC} \quad ; \quad Y_A = Y_A^{AC} \tag{4.37}$$

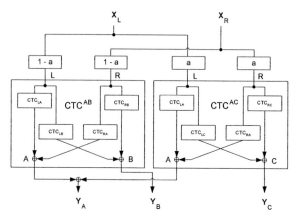

Figure 4.11: Dynamic three-channel dual CTC filter structure and cross-fading.

Hence the head-angle dependent weighting factor $a(\varphi)$ can be defined,

$$a\left(\varphi\right) = \begin{cases} 0, & \varphi \leq \varphi_0 \\ \dfrac{\varphi - \varphi_0}{\varphi_1 - \varphi_0}, & \varphi_0 < \varphi < \varphi_1 \\ 1, & \varphi \geq \varphi_1 \end{cases} \tag{4.38}$$

According to the preceding definition, the input signal is distributed to the two CTC systems,

$$X_L = \qquad\qquad (1 - a\left(\varphi\right)) \cdot X_L^{AB} + a\left(\varphi\right) \cdot X_L^{AC} \tag{4.39}$$

$$X_R = \qquad\qquad (1 - a\left(\varphi\right)) \cdot X_R^{AB} + a\left(\varphi\right) \cdot X_R^{AC} \tag{4.40}$$

Figure 4.11 shows the complete CTC filter structure of a three-loudspeaker solution. The ±45 degree and ±90 degree configurations alternate to provide full head rotation, in the same manner as practiced by the switching method. Only between two sectors both configurations are active at the same time. As mentioned above, the filters are truncated using a window of 1024 samples. The center of the window is defined by the arithmetical mean value of the location of the maximum amplitudes of all the four filters. This center is being shifted as depicted in Figure 4.12, by changing the loudspeaker configuration. The window's center position is locked to the filter set currently in use (see Figure 4.12 (a)) and is shifted for both filter sets to the center position of the succeeding filter-set (see Figure 4.12 (b))during the transition between two areas to avoid audible artifacts. In this way the current binaural audio signal is filtered with the correct CTC filter for the specific position. Initial listening experiments using this dual CTC showed that clicks are apparently no longer audible. Thus, an efficient CTC system can be established in the full space around the listener with this method.

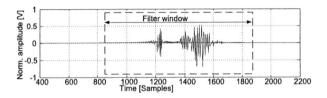

(a) ±45 deg. speaker setup

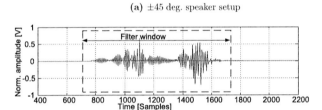

(b) ±90 deg. speaker setup

Figure 4.12: Windowing when using the three loudspeaker setup. The window position is locked to the filter set currently in use and is shifted to the succeeding filter set during the transition between two areas.

4.2.6   Loudspeaker Directivity

The directivity of the loudspeakers has not been taken into account so far. The frequency characteristic of the loudspeaker which is chosen for the reproduction (*Klein&Hummel O110D*) is almost ideally flat on-axis ($\varphi = 0$, $\vartheta = 0$) which can be seen in Figure 4.13.

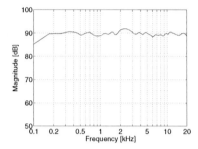

Figure 4.13: Measurement of the reproduction loudspeaker *Klein&Hummel O110D* on axis ($\varphi = 0$, $\vartheta = 0$).

Apart from the fact that the frequency characteristic should be flat, it is even more important that the frequency characteristic is identical for both loudspeakers. Unfortunately the frequency response depends on the orientation of the loudspeaker. Figure 4.14 shows the directivity of the chosen loudspeaker for the horizontal and vertical plane. Each off-axis frequency response is divided by the on-axis frequency response to obtain the relative deviation to the reference direction. It can be seen that the angle of constant radiation (drop-off below 3 dB) constitutes approximately ± 20 deg. for the horizontal (azimuth) and ± 10 deg. for the vertical (elevation) angle. The effect is even higher if the loudspeakers are located in a different height as the ears of the listener due to the higher variability of the directivity in terms of vertical angles.

In a dynamic system the relative angle between loudspeaker and listener is not fixed. Thus, the crosstalk cancellation is not able to achieve the maximal accessible channel separation, without taking the loudspeaker directivity into account.

Figure 4.15 depicts the effect which occurs when the position of the listener is not located in the center of a symmetrical loudspeaker setup. In this example, the head is still on axis with the left loudspeaker but the angle between the frontal direction of the right loudspeaker and the head has been raised to 27 deg.. If the directivity is not taken into account, the magnitude of the difference between the frequency responses of the loudspeakers corresponds to the reduction of the channel separation.

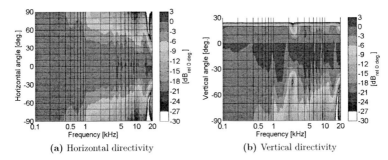

(a) Horizontal directivity   (b) Vertical directivity

Figure 4.14: Directivity plot of the reproduction loudspeaker used for the CTC system.

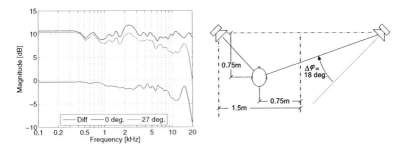

Figure 4.15: Variation of the frequency response of the loudspeaker as a function of the head position. The frequency responses of the loudspeaker are shifted by 10 dB to fit the figure.

Chapter 5

Evaluation

There are several issues that may cause errors in terms of the localization of sounds sources represented by a binaural signal:

- Insufficient channel separation during reproduction

- Frequency characteristics of the reproduction system

- Differences between HRTFs used for the synthesis or recording and listener HRTFs

- Time or phase shifting between left and right channel during reproduction

- No possibility to validate the sound source position due to head related, not room related, binaural representation.

The evaluation of the reproduction system is separated into two parts, the analytical evaluation consisting of measurements and a subjective analysis involving listening tests. The subjective tests are required for the evaluation of the performance in an acoustically deficiently environment, i.e. a not ideal absorbing enclosure which is, however, the most common field of application.

5.1   Measurements

The measurements are carried out in the semi-anechoic chamber at the Institute of Technical Acoustics. Only the direct response is analyzed. The reflections are truncated in all measurements. The measurements are evaluated in the frequency domain which means a representation of the channel separation as a function of frequency. The drawback of this representation (which is still most significant) is the fact that the frequency response is calculated by evaluating a dedicated time interval not a single "snapshot". The analysis of the complete impulse response including the interfering reflection leads to results which do not represent the perception. The ear (brain) is still able -due to the temporal displacement- to distinguish between the first arriving sound and the reflections.

The first arriving sound (*law of the first wavefront, precedence effect, Haas effect*) is considered as the most important part regarding the localization of sound. More information can be found in [Gar68].

5.1.1 Setup

The measurements are carried out at several positions and orientations of the artificial head to give an overview of the channel separation within the operation area of the system. The positions of the measurement series are variated in two different step sizes. For the evaluation of the crosstalk cancellation inside the range of the listening area the step size is defined to 15 cm. The crosstalk cancellation is also studied with a small step size of 1 cm and 1 deg. to determine the variability of the channel separation between consecutive filters. Furthermore two different positions regarding the height (1.7 m and 3 m) of the loudspeakers, and the different loudspeaker configurations are tested.

The channel separation is determined by applying the setup described in Figure 5.1. The binaural excitation signal with a length of 743 ms ($\hat{=}$ order 15, 32768 samples) contains a sweep in its left channel whereas the right channel is set to zero (see *Excitation* and *CTC* block in Figure 5.1).

If the right channel is set to zero a signal at the output of the cancellation network is produced that contains in its left channel the signal compensating the influence of the transmission to left ear. The right channel contains the compensation for the crosstalk from the left loudspeaker to the right ear. Ear means in this context the reference point defined by the microphone positions during the HRTF measurement. This reference point is regarding the ITA artificial head which is used for the measurements the entrance of the ear canal.

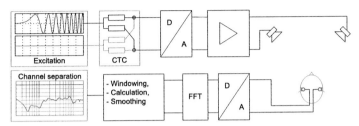

Figure 5.1: Measurement setup for the determination of the phase alignment of the cancellation filters and the resulting channel separation. As input signal a left channel sweep and right channel zero signal is used.

Two time signals, namely the crosstalk and the cancellation signal can be obtained separately at the right ear by splitting the measurements into two parts. Only one speaker is active during each of the measurement cycles. This separation is useful to determine the exact phase alignment of the compensation filters and to detect possible inaccuracies of the tracking sensor position. Figure 5.2 illustrates the signal paths and the result of the two measurement cycles (each with one active speaker) at the right ear of the artificial head. It can be seen that the signal being related to the compensation path matches very well with the crosstalk signal, but with the desired 180 deg. phase shift.

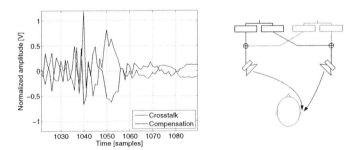

Figure 5.2: Result of the two measurement cycles (each with one speaker active) at the right ear of the artificial head.

All results of the measurements which are described in the following sections are plotted as a normalized representation. The curves are normalized to the frequency response measured at the ipsilateral (here left) ear related to the loudspeaker (here also left) which reproduces the sweep signal. The difference between the sound pressure level at the ipsilateral ear and the sound pressure level at the contralateral ear is defined as channel separation. The sound pressure level at the contralateral ear is the result of the superposition of crosstalk and cancellation signal. Furthermore, all plots are smoothed with a bandwidth of 1/3 octave. Figure 5.3 shows an example of one measurement in the absolute and in the normalized representation.

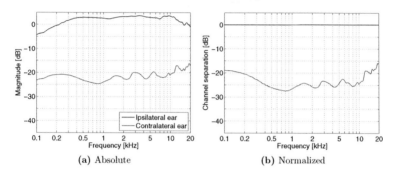

Figure 5.3: Absolute and normalized representation of the sound pressure level (magnitude) measured at the ears of the artificial head.

5.1.2 Baseline

Measurements of a static system is made to obtain a realistic reference for the channel separation, to classify the performance that could theoretically be achieved by the dynamic system. Under absolute ideal circumstances the HRTFs used to calculate the crosstalk cancellation filters are the same as during reproduction (individual HRTF of the listener). In a first test the crosstalk cancellation filters were processed with HRTFs of an artificial head in a fixed position. The windowing to a certain filter length and the smoothing give rise to a limitation of the channel separation. The internal filter calculation length is 2048 samples and for reasons discussed in Chapter 4 *Crosstalk Cancellation* the HRTFs are smoothed to reduce the small dips. After the calculation the filter set is truncated to the final filter length of 1024 samples. The dynamic system works with the same length.

The calculated channel separation using this (truncated) filter set and the smoothed HRTFs as reference is plotted in Figure 5.4 curve (a). Afterwards the

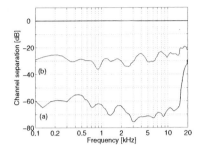

Figure 5.4: Accessible channel separation using a filter length of 1024 taps. (a) = calculated, (b) = static solution.

achieved channel separation is measured at the ears of the artificial head, which had not been moved since the HRTF measurement (Figure 5.4 curve (b)). In contrast to the ideal reference cases the following section deals with the channel separation which is achieved by the dynamic cancellation system using the same filter length but the HRTF database instead of direct measurements.

5.1.3 Results

The following figures illustrate the "global" crosstalk cancellation ability of the system within the listening area. The height of the ears is set to 1.7 m for all measurements.

Figure 5.5 depicts the results achieved during a position variation of the artificial head along the z-axis (a) and the x-axis (b) respectively. The step size is chosen to 15 cm. The maximum deviation to the center defined by the tracking transmitter is 0.6 m. [1]

It can be seen that the achieved channel separation is on average located at approximately -20 dB which is $5 - 10$ dB below the anchor defined by the static setup (see Figure 5.4). The curves show slightly different characteristics in the mid frequency range and some more distinctive deviations in the high frequency range. But nevertheless this result is excellent given a dynamic cancellation system with all the error prone variables such as calculated alignment of the HRTFs based on position measurements.

---

[1]This limitation is necessary due to increasing inaccuracies of the position measurement. This is not a fundamental problem of the method, but a technical limitation in the test environment. Using a tracking device which is not based on electro-magnetical measurements would not cause this limitation.

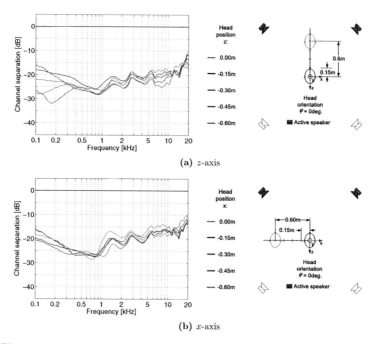

Figure 5.5: Channel separation measured for the variation of the head position along the z-axis and the x-axis respectively. Loudspeakers are mounted at 1.70 m.

A combination of both displacements is shown in Figure 5.6. The measurement described in Figure 5.5(b) is repeated, however, with a constant offset in the z-direction of -0.6 m to further evaluate the lateral positions (along the x-axis) which are more critical than a displacement along the z-direction only. The angles between head and loudspeakers are varying more from one position to the next with this setup, than in the previous measurements, due to the closer distance to the loudspeakers. The channel separation remains nearly constant in this area. Only the last curve shows some deviation which can be explained by the inaccuracies of the position measurement.

In a next step the channel separation achieved by using the ±90 degree loudspeaker configuration is measured. This configuration is activated automatically when the listener is facing one of the other loudspeakers. For this reason the orientation of the artificial head is set to 45 degree. Due to the symmetry of the setup only the variation

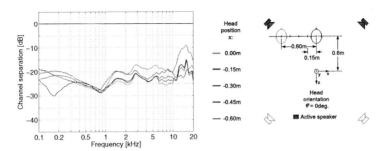

Figure 5.6: Channel separation measured for the variation of the head position along the x-axis, but with a constant offset regarding the z-axis of -0.6 m. Loudspeakers are mounted at 1.70 m.

along the x-axis is evaluated. The results are depicted in Figure 5.7.

The channel separation is in the same range as in the previous measurements. Furthermore, the performance at higher frequencies is even better, which can be explained by the head shadow. The head shadow is most significant at high frequencies for lateral sound incidence which occurs when this loudspeaker configuration is used.

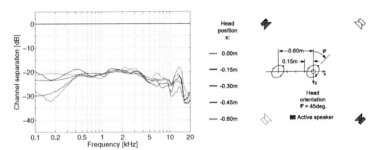

Figure 5.7: Channel separation measured for the variation of the head position along the x-axis and a fixed head rotation of $\varphi = 45$ deg.. Loudspeakers are mounted at 1.70 m.

The channel separation is evaluated so far by measurements at different positions which are at least 15 cm apart from each other. The results show that the differences between some of the curves exceed several dB, which might be audible during the movement of the listener. For this reason comparing measurements carried out at positions with a closer spacing is an interesting and important examination to determine the variability of the channel separation between consecutive filters. The closer spacing corresponds to a greater extent to a natural movement of a listener.

Figure 5.8 depicts a series of measurements with a step size of 1 cm along the x-axis (a) and a constant offset in the z-direction of 0.6 m (b). The differences of the curves among each other are remarkably small. It can be assumed -without any subjective evaluation- that the variation of the channel separation will not cause any audible artifacts if the filter is updated fast enough. This assumption has to be validated later on.

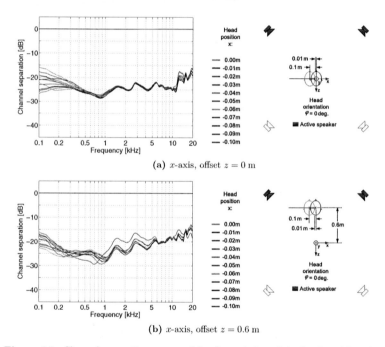

(a) x-axis, offset $z = 0$ m

(b) x-axis, offset $z = 0.6$ m

Figure 5.8: Channel separation measured for the variation of the head position along the x-axis with different offsets in z-direction. Loudspeakers are mounted at 1.70 m.

While studying the *sweet spot* the rotation of the head about the y-axis was identified as critical as the dimension of the sweet spot is very small. For this reason the channel separation is measured for both loudspeaker configurations each in a range of 10 deg. with a step size of just 1 degree.

Figure 5.9b(a) shows the results of the measurements between 0 and 10 deg. which

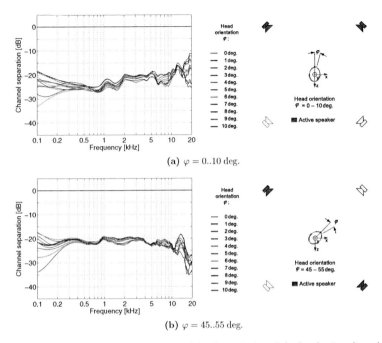

(a) $\varphi = 0..10$ deg.

(b) $\varphi = 45..55$ deg.

Figure 5.9: Channel separation measured for the variation of the head azimuth angle (φ). Loudspeakers are mounted at 1.70 m.

means that the ±45 deg. loudspeaker configuration is used. The ±90 deg. loudspeaker configuration is active during the measurements between 45 and 55 deg. depicted in Figure 5.9b(b). The plots show deviations that are slightly larger than those for a lateral movement, but they are still very good for this critical movement. It will have to be evaluated perceptually whether this variability of the channel separation is still low enough for a qualitative sound reproduction or not (see Chapter 7).

The crosstalk cancellation performance of the system is evaluated separately for both loudspeaker configurations so far. But as described in Section 4.2.5 *Dual Crosstalk Cancellation* it is necessary to implement a fading between both configurations to guarantee a smooth sector changeover. Within a small area both configurations are active and an angle dependent fading is applied. In this case the compensation signal consists of a superposition of the contribution of each loudspeaker.

This means for the test environment used here the two loudspeakers of the right hand side of the artificial head. The complete cancellation process can be described as a constructive superposition of the two compensation signals and a destructive superposition of the crosstalk.

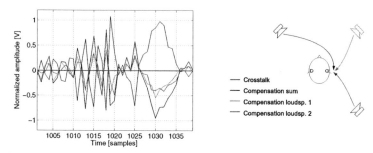

Figure 5.10: Zoomed time plot of the cancellation signal superposition and the alignment to the crosstalk signal which has to be canceled.

Figure 5.11 illustrates the superposition of the compensation signal which is phase inverted to the crosstalk signal. It should be noted that due to the additional component that has to be aligned exactly, this dual crosstalk cancellation is more error-prone than the two channel configuration.

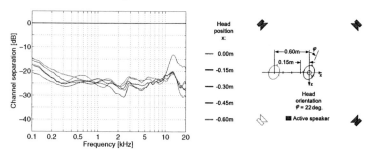

Figure 5.11: Channel separation measured for the variation of the head position along the x-axis. The head azimuth angle is set to $\varphi = 22$ deg. to activate the dual crosstalk cancellation. Loudspeakers are mounted at 1.70 m

It can be seen that the results are very similar to the channel separation achieved by means of the stereo crosstalk cancellation. This shows that the setup is calibrated

with sufficient accuracy (positions of the loudspeakers). Furthermore it proves that
the dual crosstalk cancellation presented in this thesis is an appropriate method to
combine the two stereo crosstalk cancellation system to cover a complete 360 degree
rotation of a listener.

Finally, the loudspeakers are mounted at a height of 3.0 m and the measurements
are repeated for the variation of the position along the x-axis and the z-axis respectively. This is of special interest as the speakers will be mounted at the same height
in the *CAVE-like* environment. Figure 5.12 shows the results.

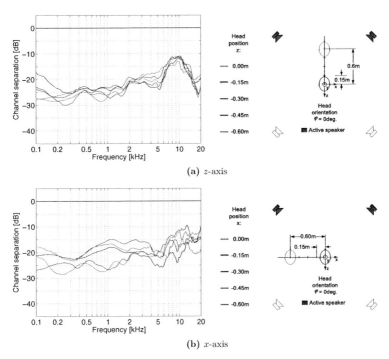

(a) z-axis

(b) x-axis

Figure 5.12: Channel separation measured for the variation of the head position along
the z-axis and the x-axis respectively. Loudspeakers are mounted at 3.0 m.

The achieved channel separation is slightly lower and the spreading of the results
is higher. This can be explained by the decreased ILD of the HRTFs which have
to be used for the filter generation. The ILD as well as the ITD are lower for

HRTFs defining the transfer function of an elevated source to the ears. However, the channel separation achieved by the dynamic system is still excellent. In comparison to the baseline (see Figure 5.4) which was determined under ideal conditions (direct measurement, no movement of the head between filter calculation and evaluation) the channel separation has not decreased as much as expected by using the dynamic solution.

5.2   Influence of Misalignment

All measurements described in the section above deal with fixed listening (measurement) positions. A movement during the measurement is not considered. Using the correlation measurement method a movement of source or receiver during the recording of the signal is not permitted as it destroys the postulation of a Linear Time Invariant (LTI) system. There are a number of influences which can have an effect on the achieved channel separation of the dynamic system, e.g. tracking accuracy, tracking latency, calculation time of the filters, etc.. In this section the influence of a filter misalignment caused by any of the reasons mentioned above is evaluated. The channel separation is determined at several closely spaced positions before a filter update is performed.

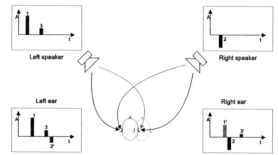

Figure 5.13: Misalignment of the head position related to the position the filter is calculated for.

The dimension of the *sweet spot* which defines the area where no changes were detected has been evaluated in Section 4.1.1. A dynamic adaption of the filter is realized as a changeover from one discrete *sweet spot* to the next. The fact that one filter is switched or faded to the next denotes that the listener can immediately compare the result generated by the old filter with the result generated by the new filter. While evaluating a static setup the coloration or location of a sound changes continuously by moving the head. However, an abrupt change to a new filter does not exist. A measurement series is carried out to determine the channel separation while moving the artificial head in small steps of 5 mm suppressing any filter update of the CTC system. The result is plotted in Figure 5.14(a) over a frequency range of 100 Hz to 20 kHz. Each of the curves represents a single position. The channel separation is decreasing with every step over a wide range up to the state at which the cancellation has nearly no effect except of the frequency range around 8 kHz. This region seems to be not affected by the misalignment which can be explained by the natural shading of the head in this frequency range for the specific angle to the loudspeaker. The channel which represents the contralateral ear of a HRTF measured at 45 deg. shows a very large attenuation at this frequency range (see Figure 5.14(b)).

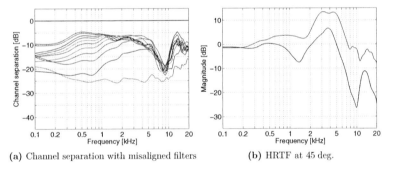

(a) Channel separation with misaligned filters   (b) HRTF at 45 deg.

Figure 5.14: Measurement results of the channel separation for misalignment of the real head position and the position the filter is being calculated for (a) and the transfer function from one loudspeaker to the ears (b).

Figure 5.15 depicts the results for a series of measurements where the filter calculation is triggered manually at different positions of the head. The channel separation is plotted successively as a function of the head position for different single frequencies. It can be seen that the channel separation decreases which implicates that the level at the contralateral ear is rising. Assigning a filter update at a certain position causes an immediate rise of the channel separation and also a reduction of the crosstalk signal. This behavior has to be kept in mind for the design of the complete audio reproduction system. It is very important that the filter update can be performed fast enough to avoid large differences regarding the channel separation. The inversion of this argument is that the listener is not allowed moving too fast.

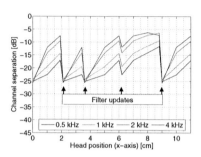

Figure 5.15: Channel separation as a function of the head position for different frequencies. The position of the artificial head is variated along the x-axis with a step size of 5 mm. The filter update has been triggered manually at the assigned positions.

5.3 Influence of Reflections

The theory of canceling the crosstalk is based on the assumption of a reproduction in an anechoic environment. However, the system discussed here is supposed to work together with visual VR systems such as a *CAVE-like* environment or a *L-Bench*. The projection walls of such environments usually consist of solid material which cause interfering reflections as can be seen in Figure 5.3. These reflections leads to a decrease in performance of the CTC system and may reduce the ability to localize the virtual sources in the correct position.

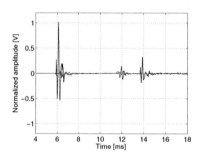
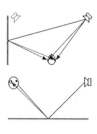

Figure 5.16: Binaural impulse response with reflections.

For that reason the influence of reflections has to be examined to ensure the system's applicability. A subjective listening test is carried out in the semi-anechoic chamber of the ITA with three additional walls of plywood to simulate the video screens in order to analyze the differences in localization by comparing an ideal environment to an environment with reflecting walls around the test person. The test is performed in an anechoic, not in the real, environment to get direct comparable results just by adding the reflecting walls in an otherwise identical setup [Len03]. A complete analysis with a localization test in a *CAVE-like* environment is described later in Chapter 7 *Validation*, when the complete spatial audio platform has been introduced. However, this test is not only valid with regard to *CAVE-like* environments, but also universally.

The listening test is performed by 15 subjects, the answers are recorded by the experimenter. A pulsed pink noise of 200 ms duration followed by 500 ms of silence is used as stimulus. Twelve directions (every 30 deg.) are tested in the horizontal plane, repeated three times in a randomized sequence. The subjects are asked to name the direction of the virtual source relative to their line of sight. The statements are reported to the experimenter in analogy to a clock dial, e.g. twelve o'clock for a source in front and tree o'clock for a source on the right hand side of the listener. As representation of the results scatter plots seems to be adequate because the range and the type of deviation as well as the amount can be depicted simultaneous. All trials which have been perceived correctly are located on the diagonal of the plot. The diameter of a circle represents the number of trials that caused the specific mapping of real and perceived positions.

For testing the benefit of the crosstalk cancellation system the first test is a simple setup where the preprocessed binaural signal is distributed directly by the stereo loudspeaker set without any signal processing during runtime. In this case all spatial information is related firmly to the listener's head and also turns with the head. The scatter plot in shows the perceived angles plotted versus the presented angle. The "sine" shaped arrangement (see Figure 5.17) of the answers of the subjects around the 0 deg. stimulus / 0 deg. perception point illustrates the inability of the arrangement to place virtual sources in the region behind the frontal plane. The crosstalk destroys the temporal and frequency related cues which are essential to perceive sources from behind.

In contrast to the pure reproduction of the binaural signal over the loudspeakers,

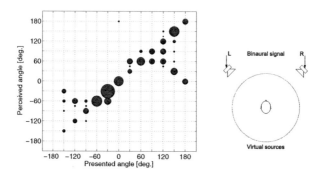

Figure 5.17: Listening test results and test environment. The binaural signal is distributed directly by the loudspeakers without any crosstalk cancellation.

the next test is carried out with the dynamic CTC system while the spatial information is still head related. The results are plotted in Figure 5.18. The arrangement of the answers along the diagonal of the scatter plot indicates that most of the listeners perceive the virtual source in the supposed direction. Together with the baseline test without any crosstalk compensation (Figure 5.17) this result shows the significant influence of the crosstalk compensation for an enhancement of the localization accuracy.

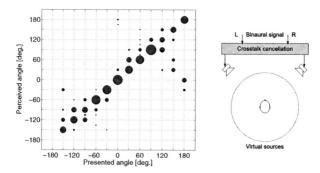

Figure 5.18: Listening test results and test environment. The binaural signal is distributed over loudspeaker after passing a crosstalk cancellation filter.

The next test is carried out with three additional reflecting walls around the listener. The chosen material provides roughly the same acoustical properties as the fiberglass used for the video projection. The test environment has nearly the same dimensions as the *CAVE-like* environment and is a good substitution to examine the ability of virtual source localization in reverberant environments.

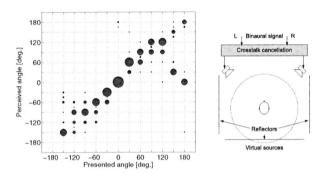

Figure 5.19: Listening test results and test environment. The environment is not ideal anechoic. The binaural signal at the ears is crosstalk compensated.

The localization results in an anechoic chamber, an ideal acoustical environment, are excellent but by adding the reflecting walls to the listening environment the performance decreases significantly. Figure 5.19 shows that more front-back confusions occur as well as an increasing inaccuracy for lateral virtual sources. The localization performance achieved here differs significantly from the results achieved in the complete anechoic environment. It should be noticed that the dynamic binaural synthesis is still not active in this test. This means that the listener has no ability to improve the location of the virtual source by slight head movement. As described in Section 3.3 *Dynamic Aspects* in Chapter 3 *Binaural Synthesis* this fact is the major advantage of a dynamic synthesis.

During the last listening test the synthesis is performed on-line (dynamic synthesis), and the relative direction of the source is updated instantly. This generates a virtual source located at a fixed position in the room. Now the listener can turn the head towards the direction where the source position is assumed to approve the impression.

The results shown in Figure 5.20 are very good despite the reflections and they reveal a significant improvement of the localization. In this test only a dynamic two-loudspeaker CTC system is used so that the listeners could not turn completely

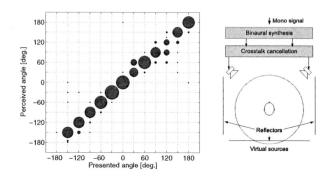

Figure 5.20: Listening test results and test environment. The environment is not ideal anechoic. The binaural signal is generated by the dynamic binaural synthesis to realize room-related virtual sources. The signal at the ears is crosstalk compensated.

towards most of the virtual sources. However, most of the sources are perceived in their designated directions (circles on the diagonal). Only very few front-back confusions occur and the number and the range of the deviations are smaller. Since the reflections arrive at the ears later than the direct signals, humans are still able to detect the right direction of the source. These tests show that dynamic binaural synthesis together with dynamic CTC is an appropriate technique for sound reproduction in VR systems.

Tests of other labs and different CTC systems indicate a better subjective performance than it would be expected from measurements including all reflections as well. An extensive study about the performance of crosstalk cancellation systems in not ideal anechoic environments can be found in [War01] and [TNKH97]. One aspect which confirms this phenomenon is the precedence effect by which sound localization is primarily determined by the first arriving sound [DC06]. An other aspect is the head movement which gives the user the ability to approve the perceived direction of incidence.

Chapter 6

Interactive VR-System

The main focus of this thesis is to establish a system for the auralization of arbitrary virtual sound scenes. A sound scene is described by a set of parameters such as the position of each source, the individual directivity, the source content, the listener position, and the acoustical attributes of the simulated environment. Furthermore, the parameter set does not remain static. It can be updated dynamically by the system as well as by the user, either by direct or indirect interaction. The techniques described in the preceding chapters are used to realize a system according to the depicted constrains, especially the dynamic adaption on the variation of any parameter in real-time. The integration of the aural component into a visual Virtual Reality system enables the creation of complex multi-modal virtual environment with a high degree of naturalness.

6.1 Technical environment

The complete VR system consists of a number of different subsystems. Only the cooperation of all these subsystems enables the generation of complex and in this case multi-modal virtual scenes. The main application has to handle all required information, the temporal flow of the presented content, the associated data, and has to distribute these data to the relevant subsystems. This application is running on the same platform which is responsible for the visual rendering.

Furthermore, the acoustical attributes of the virtual sound field have to be provided to the audio rendering server. Due to the high complexity of the sound field modeling a dedicated subsystem is used to generate the information. In this section the two main components of the system, which provide input data for the audio rendering, are described.

The test system which is used to discuss all further results regarding processing time consists of the following components: As a visual VR machine, a dual Pentium 4 machine with 3 GHz CPU speed and 2 GB of RAM is used (cluster master). The host for the audio VR subsystem is a dual Opteron machine with 2 GHz CPU speed and 1 GB of RAM. As audio hardware an RME Hammerfall system is used [RME]. The RME hardware allows the sound output streaming with a scalable buffer size and therefore a minimum latency of 1.5 ms. The network interconnection between the machines a standard Gigabit Ethernet is used. The room acoustical simulations run on Athlon 3000+ machines with 2 GB of RAM.

6.1.1   Visual VR-System

The task of the visual VR system is to generate a three-dimensional visual representation of the virtual scene. In contrast to a flat or perspective two-dimensional image, the holographic video technique is used to create the illusion that the displayed objects have a volume. To achieve this, it is required to calculate images separately for each eye of the user with a perspectively correct adjustment for the specific position of each eye.

At the beginning of the visual reproduction systems development for Virtual Reality, HMDs were used to generate the stereoscopic view. This technology uses small video panels mounted at a short distance in front of each eye. The drawback of these systems is the weight of at least several kilograms and the fact that the user can not see his own body anymore, so that a synthetic representation of the users extremities is required. Room-mounted displays generate the stereoscopic images on screens and are called, depending on how many screens are used or which angle is spanned by spherical displays: *Powerwall* (one surface), *L-Bench* (two surfaces), *CAVE-like display* (> two surfaces). In such an environment the user is able to move around and see him- or herself acting in the virtual scenery. In contrast to HMDs where the images can be reproduced physically separately for each eye, room-mounted displays use the same screen for the images for both eyes. Hence, the pictures for both eyes have to be separated again to generate the holographic perception of the generated three-dimensional objects.

In analogy to the acoustical crosstalk cancellation, where the two audio channels have to be separated at the ears, an appropriate technique is required to separate the two images at the eyes. The three most popular techniques for separating two images from one projection plane are time encoding (shutter-glasses), light direction encoding (polarized filters) and color encoded (anaglyph-glasses).

The VR system primarily used for the implementation and testing of the audio reproduction system, described in this thesis, is the *CAVE-like* environment installed at the Center for Computing and Communication (CCC), RWTH Aachen University. This system is realized as a five-sided projection system of a rectangular shape (see Figure 6.1). The special shape enables the use of the full resolution of 1600 by 1200 pixels of the Liquid Crystal Display (LCD) projectors on the rectangular walls and the floor, and a 360 deg. horizontal view. The dimensions of the projection volume are $3.6 \text{ m} \times 2.7 \text{ m} \times 2.7 \text{ m}$ yielding a total projection screen area of 26.24 m^2. Additionally, light direction encoding (passive stereo) via circular polarization allows the use of light-weight glasses for the image separation. Head- and interaction device tracking is realized by an optical tracking system [ART04]. The setup of this display system is an improved implementation of the system that [CNSD93] was developed with the clear aim to minimize attachments and encumbrances in order to improve user acceptance.

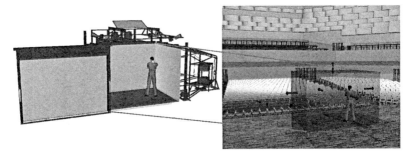

Figure 6.1: Viewing "through" the display system into the virtual world (right) and schematic draw of the *CAVE-like* display system installed at the Center for Computing and Communication (CCC), RWTH Aachen University.

CAVE-like environments enable the user to move directly in the scene, e.g. by walking inside of the boundaries of the display surfaces and tracking area. Additionally, indirect navigation enables the user to move in the scenery virtually without moving his body but by pointing metaphors when using hand sensors or joysticks. Indirect navigation is mandatory, e.g. for architectural walk-throughs as the virtual scenery is usually much larger than the space covered by the *CAVE-like* device itself. But this also implies that two different coordinate systems are required for a complete definition of the environment. On the one hand the World Coordinate System (WCS) which is related to the virtual world and on the other hand the Real Coordinate System (RCS) which is related to the physical environment namely the display system. Objects existing only in the virtual world are defined using the WCS (e.g. virtual sources), positions of objects which are only exist in the real world will be described using the RCS (e.g. the reproduction loudspeakers). Finally, one "object" has to be described simultaneously according to both coordinate systems which is the user. The user acts in the virtual world at a relative position to all virtual objects and simultaneous in the real world at a relative position to the physical system.

6.1.2   Interface to Room Acoustical Simulation

All parameters such as positions and movement of the virtual sources and the listener can so far be provided by the VR application to update the audio rendering subsystem, except for the definition of the acoustical behavior of the virtual scene itself. For this reason an interface to a room acoustical simulation system is established to enhance the complete aural representation with an adequate representation of the complex sound-field of the virtual enclosure.

To provide a system that is capable of simulating different types of room geometries, such as extremely long, flat, or at least coupled rooms, simple room acoustical effect algorithms are often considered as not sufficient. In this case, a more physically based description of the sound field is required. Geometrical acoustics is the most important model used for auralization in room acoustics [Kut00]. Wave models would be more exact, but only the approximations of geometrical acoustics and the corresponding algorithms provide a chance to simulate room impulse responses in real-time.

In geometrical acoustics, deterministic and stochastic methods are available. All deterministic simulation models used today are based on the physical model of image sources (IS) [AB79]. The concept of the image source method provides a flexible data organization. The on-line movement of primary sound sources and their corresponding image sources is supported and can be updated within milliseconds. Unfortunately, the computational costs are dominated by the exponential growth of image sources with an increasing number of polygons and reflection order. But it is also known that, from reflections of order two or three, scattering becomes a dominant effect in the temporal development of the room impulse response [Kut95] even in rooms with rather smooth surfaces. This fact requires an additional approach, as scattering is basically not supported by the image source method. Stochastic ray tracing is an approach which is feasible to calculate the scattering part required for a physically more exact representation of reverberation.

However, the image source method is appropriate to describe the early specular reflections, which are also very important for the localization of a sound source in a room. Figure 6.2 depicts the principle of the image source method. The important audibility test is the most time-critical process due to the determination of ray-wall intersections. Binary space partitioning (BSP)-trees [SBGS69] is used to organize the geometrical data of the room and thus, to speed up the intersection test [SL06]. In addition, instead of the time-critical regeneration of the image sources regarding a changed receiver's position, it is possible to translate them fast to their new positions by means of transformation matrices.

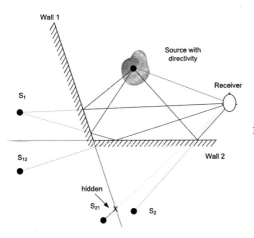

Figure 6.2: Principle of image sources. The primary source is mirrored at each wall to generate possible first order reflections. The second order reflections are represented by mirroring the first order image sources again at all walls.

The room acoustical simulation transfers a list of audible image sources to the audio server on which the final filter is processed (see Section 6.2.1). The image sources contain all relevant geometrical data as well as information about the material of the walls (absorption).

It should be kept in mind that the two basic physical methods, the deterministic image sources and the stochastic scattering, should be taken into account in a sound field model with a certain degree of realistic physical behavior. The contribution to the binaural impulse response related to the ray tracing process is calculated directly on the room acoustics server in order to reduce the amount of data which has to be transferred via network. The audio server integrates the processed contribution to the final binaural room impulse response. This part of the room acoustical simulation is a topic of current research at the Institute of Technical Acoustics and will not be discussed further in this thesis. A detailed description of the concept and further information can be found in [LSVA07, SDV07].

6.2   Audio Server

This system integrates two main parts required for a sophisticated spatial audio system. This is on the one hand the generation of the spatial distributed sound sources using the binaural synthesis extended by the directivity information of the source and information provided by the room acoustical simulation. On the other hand the aurally correct reproduction of the generated binaural signal by using the dynamic crosstalk cancellation. As stated before, the loudspeaker based CTC is used as a consequence of the postulation not to decrease the non-intrusive visual system.

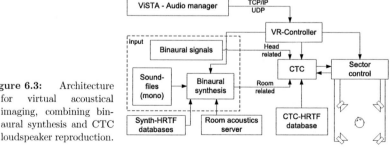

Figure 6.3:   Architecture for virtual acoustical imaging, combining binaural synthesis and CTC loudspeaker reproduction.

The general architecture and interdependence of the involved subparts is shown in Figure 6.3. The figure presents a box "ViSTA-Audio Manager" which is a substitute for an arbitrary VR software system that is capable of delivering spatial information of different sound sources and the listener's head by network based communication. But so far, ViSTA (Virtual Reality for Scientific Technical Applications) is the only toolkit which provides all information required for addressing the complete list of features of the audio system.

The system distinguishes between two types of update messages. One type deals with low frequency state changes, such as commands to play or stop a specific sound. The second type updates the spatial attributes of the sound source and the listener at a high frequency. For the first type a reliable transport protocol is used (TCP) while the latter is transmitted at high frequency over a low overhead but possibly unreliable protocol (UDP).

A more detailed system layout concerning the filter structure is shown in Figure 6.4. The synthesis section, connected to the VR-controller, which provides the position data, accomplishes the multi-track convolution of the mono sound files or sound device input channels and the down-mix of the processed binaural output. The other main part of the system is the dual CTC unit.

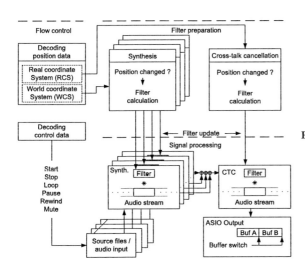

Figure 6.4: Filter structure of the audio rendering system. Binaural synthesis is updated according the world coordinate system (WCS), the CTC is updated according to the real coordinate system (RCS).

Both subsystems require different coordinate systems which were introduced above, to cover the possibility of moving the virtual world relative to the user. This means that the center of the world coordinate system has not the same location or orientation as the center of the coordinate system which is fixedly related to the physical coordinates of the display system. As all virtual sources move with the virtual world, the binaural synthesis, which is described in the following section, has to be updated according to the world coordinate system.

6.2.1   Synthesis Filter

The dynamic auralization requires that the filters can be updated very fast, which is even more challenging if the room acoustics has to be taken into account. Furthermore, the update time becomes also more important in combination with congruent video images. Thus, the filter processing is a crucial part of the real-time process.

The whole filter construction is separated into two parts. The most important section of a binaural room impulse response is the first part containing the direct sound and the early specular reflections of the room. These early reflections are represented by the calculated image sources and have to be updated at a rate which has to be sufficient for the binaural processing. Therefore, the list of the currently audible sources represents the operation interface between the room acoustics server and the audio server. The diffuse and late specular part of the room impulse response is calculated on the room acoustics server to minimize the time required by the network transfer, because the amount of data required to calculate the room impulse response is significantly higher than the resulting filter itself.

Every single fraction of the complete impulse response, either the direct sound or the sound reflected by one or more walls, runs through several filter elements, as shown in Figure 6.5. Elements such as directivity, wall, and air absorption, are filters in a logarithmic frequency representation with a third octave band scale with 31 values ranging from 20 Hz to 20 kHz. These filters contain no phase information so that only a single multiplication is needed for each band. The drawback of using a logarithmic representation is the necessity of interpolation to multiply the resulting filter with the HRTF. But this is still not as computationally expensive as using a linear representation for all elements, particularly, if more wall filters have to be taken into account for the specific reflection.

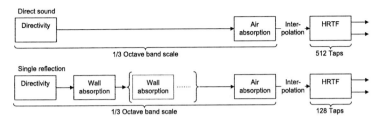

Figure 6.5: Filter elements for direct sound and reflections.

The first filter element is the directivity which has an important influence on the sound in a room. The directivity, describing the direction-dependent radiation of the source, defines the weighting factor for every single contribution (direct sound or a reflection) to the complete impulse response. This is even more important for a dynamic auralization, where not only the listener is able to move and interact with the scenery, but where the sources can also move or turn. The user expects a different sound coloration if he is moving in front of or behind a source with a directional radiation, e.g. a trumpet. The naturalness of the whole generated sound scene is improved by every dynamic aspect being taken into account. The program accepts external directivity databases of any spatial resolution. The internal database has a spatial resolution of 5 degree for azimuth and elevation angles. This database contains the directivity of a singer and several natural instruments as described in Chapter 2. Furthermore, it is possible to generate the directivity manually.

So far, the wall absorption filters are independent of the angle of sound incidence, which is a common assumption for room acoustical models. It can be extended to consider angle-dependent data, if necessary. Reflections calculated by using the image source model will be attenuated by the factor of the energy, which is distributed by the diffuse reflections. The air absorption filter is only distance-dependent and is applied also to the direct sound, which is essential for huge distances between the listener and source.

At the end of every filter pass, which represents, up to now, a mono signal, an HRTF has to be used to generate a binaural head related signal, which contains all directional information (see Chapter 3). For the direct sound, an HRTF of a higher resolution is used to ensure a sufficient amount of directional information of the sound source. The FIR filter length is chosen to 512 taps. Due to the fact that the filter processing is done in the frequency domain, the filter is presented by 257 complex frequency domain values corresponding to a linear resolution of 86 Hz. Distances in the range of 0.2 m to 2 m between the source and the listener are realized by using the near-field HRTFs described in Section 3.2. The spatial resolution of the HRTF databases is 1 degree for azimuth and 5 degree for elevation angles for both the direct sound and the reflections.

The length of the HRTFs used for the contribution of image sources is lower than for the direct sound, i.e. 128 taps which corresponds to 65 complex values, respectively, in the frequency domain. Using 128 FIR-coefficients still leads to sufficient localization results but causes a considerable reduction of the processing time (see Table 6.1). Studies regarding the effects of reduced filter length on localization can be found in [KC98]. The spatial representation of image sources is realized by using HRTFs measured in 2.0 m. In this case, this does not mean any simplification because the room acoustical simulation using image sources is not valid anyway at distances close (a few wavelengths) to a wall. A more detailed study concerning this topic can be found in [RV03] and [SN97].

Finally, after the calculation of the image source filter the audio server has to combine this filter with all filter segments of the reverberation tail (see Figure 6.6).

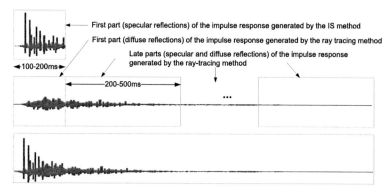

Figure 6.6: Combination of filter (or filter segments) for one ear generated by ray tracing and the first part of the impulse response generated by the image source model.

Table 6.1 depicts the processing time for the single filter fragments to give an overview of the processing time being required for the filter calculation. It should be mentioned that the length of the reverberation is room-dependent. Thus, the complete sum of the processing time is dependent on the room geometry, the number and positions of the primary sources, and the position of the listener. All measurements are performed on the test system described above.

Processing step	Time
Direct sound (512 taps)	300 µs
Single reflection (aver.)	50 µs
Preparation for segmented convolution	1.1 ms

Table 6.1: Calculation time of several parts of the filter.

6.2.2 Low Latency Convolution

Part of the complete dynamic auralization system requiring a high amount of processing power is the convolution of the audio signal. When the auralization of a room is desired instead of a free-field representation, the filter length is rising from just a few hundred taps to maybe several hundred thousand depending on the complete reverberation time of the room. A concert hall, for example, has a typical reverberation time of 1.8 s to 2.3 s which corresponds to a filter length of ≈ 80,000 taps to ≈ 100,000 taps, assuming a sample rate of 44.1 kHz. The reverberation time of a church is sometimes higher than 10 s (> 400,000 taps).

A pure FIR filtering would cause no additional latency, except for the delay of the first impulse of the filter. However, it causes the highest amount of processing power as well. Impulse responses of 100,000 taps or more cannot be processed in real-time on a PC-system using FIR filters in the time domain. The block convolution is a method that reduces the computational cost to a minimum, but the latency increases in proportion to the filter length. The only way to minimize the latency of the convolution is a special conditioning of the complete impulse response in filter blocks. Basically, the algorithm used here, works in the frequency domain with small block sizes at the beginning of the filter and increasing sizes to the end of the filter. More general details about these convolution techniques can be found in [Gar95]. The

performance of a segmented FIR convolution engine is highly affected by the way
the FIR filter is partitioned. Hence, the algorithm used here does not operate on a
segmentation that doubles the block length every other block, which is commonly used
(see Figure 6.7). The system provides a special block size conditioning with regard to
the specific PC-hardware properties as, for instance, cache size or special processing
structures such as SIMD (Single Instruction Multiple Data). Considerations which
have been made so far [Gar02, Gar68] were based on calculations of the analytic
algorithmic complexity (number of multiplys/adds).

Figure 6.7: Example of a classical segmentation of the low latency convolution.

The segmented convolution adds a time delay of only the first block to the latency
of the system. Thus, it is recommended to use a block length as small as possible.
The amount of processing power is not linear to the overall filter length and also con-
strained by the chosen start block length. Due to this fact, measurements have been
carried out to determine the processor load of different modes of operation. The test
system depicted above is used for these measurements as well. The segmentation used
currently in the system is "hand-optimized" for the specific hardware. An example of
the achieved performance is listed in Table 6.2.

Filter length	Number of sources							
	3	10	15	20	3	10	15	20
	(Latency 256 taps)				(Latency 512 taps)			
0.5 s	9%	30%	50%	76%	8%	22%	30%	50%
1.0 s	14%	40%	66%	-	11%	33%	53%	80%
2.0 s	15%	50%	74%	-	14%	42%	71%	-
3.0 s	18%	62%	-	-	16%	53%	-	-
5.0 s	20%	68%	-	-	18%	59%	-	-
10.0 s	24%	-	-	-	20%	68%	-	-

Table 6.2: CPU load of the low latency convolution algorithm.

The realization of an automated determination of the segmentation which fits best
to the used hardware is also a topic of current research at the CCC and the ITA.

The filter partitioning should be optimized based on recorded benchmark data of the specific system. Besides the lowest total computation cost, other optimization criteria, like real-time stability and an even distribution of the computational load over the time, is currently investigated [Wef07].

6.2.3 Crosstalk Cancellation Filter

The CTC subsystem is updated with position data according to the real coordinate system, because the algorithm requires the position and orientation of the head relative to the positions of the reproduction loudspeakers which are related to the physical display system. Figure 6.8 depicts the loudspeaker setup, mounted at the ceiling of the display system.

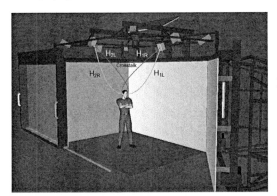

Figure 6.8: Schematic draw of the open *CAVE-like* display system with the loudspeaker used for the crosstalk cancellation installed at the ceiling. The arrows depict the signal paths to the user's ears

Each time the head position and orientation is updated in the system, the deviation of the head to the condition which caused the preceding filter change is calculated. Every degree of freedom is weighted with its own factor and then summed up (see Equation 6.1). Thus, the threshold can be parameterized in six degrees of freedom, positional values $(\Delta x, \Delta y, \Delta z)$ and rotational values $(\Delta \varphi, \Delta \vartheta, \Delta \rho)$. A filter update will be performed when the weighted sum is above 1.

$$s = \frac{|x_{new} - x_{old}|}{\Delta x} + \frac{|y_{new} - y_{old}|}{\Delta y} + \frac{|z_{new} - z_{old}|}{\Delta z}$$
$$+ \frac{|\varphi_{new} - \varphi_{old}|}{\Delta \phi} + \frac{|\vartheta_{new} - \vartheta_{old}|}{\Delta \vartheta} + \frac{|\rho_{new} - \rho_{old}|}{\Delta \rho} \geq 1 \quad (6.1)$$

The lateral movement and the head rotation in the horizontal plane are most critical so $\Delta x = \Delta z = 1$ cm and $\Delta \varphi = 1$ deg. are chosen to dominate the filter update ($\Delta y = 3$ cm, $\Delta \vartheta = \Delta \rho = 3$ deg.). The threshold always refers to the value where the limit was exceeded for the last time. The resulting hysteresis prevents a permanent switching between two filters as it may occur when a fixed spacing determines the boundaries between two filters and the tracking data jitter slightly.

One of the fundamental requirements of the sound output device is that the channels work absolutely synchronously. Otherwise, the calculated crosstalk paths do not fit with the given conditions. On this account, the special audio protocol ASIO designed by Steinberg for professional audio recording is chosen to address the output device [Ste04].

The latency of the audio reproduction system can be defined as the time elapsed between the update of a new position and orientation of the listener, and the point in time at which the output signal is generated with the recalculated filters. The output block length of the convolution (overlap save) is 256 taps as well as the chosen buffer length of the sound output device. This leads to a time between two buffer switches of 5.8 ms at 44.1 kHz sampling rate for the rendering of a single block. The calculation of a new CTC filter set (1024 taps) takes 3.5 ms on the test system. In a worst case scenario, the filter calculation is just finished after the sound output device has fetched the next block. So it takes the time while playing this block until the updated filter becomes active at the output. This causes a latency of one block. In such a case, the overall latency accumulates to 9.3 ms.

6.2.4   Communication

The interface between the visual VR system and the audio server consists of two bidirectional (TCP) and two unidirectional (UDP) communication channels. The first TCP channel establishes the connection to the audio server and allows to control the sound system. The second TCP channel is automatically created by the system and is used for server sided events, errors and exception messages to the VR client application. The UDP channels exist for the fast rate transmission of spatial updates of various sound sources in the virtual environment and the listener, encoded as a table of positions and orientations. The spatial updates are used to re-calculate the filters on-line. In contrast to this, the TCP channels are expected to be used at a low update frequency, usually related to the frame-rate of the video image rendering.

In order to obtain an estimation of the costs of network transport, the largest possible TCP and UDP messages produced by the system is transmitted from the VR application to the audio server many times and then sent back. The transmission time for this round-trip was taken and cut in half for a single-trip measurement. The worst case times of the single-trips are taken as a basis for the estimation of the overall cost introduced by the network communication. The mean time for transmitting a TCP command is 0.15 ms ± 0.02 ms. The worst case transmission time on the TCP channel is close to 1.2 ms. UDP communication was measured for 20000 spatial update tables for 25 sound sources, resulting in a transmission time for the table of 0.26 ms ± 0.01 ms. It seems surprising that UDP communication is more expensive than TCP, but this can be put down to the larger packet sizes of an spatial update ($\approx$ 1kB) in comparison to small TCP command sizes ($\approx$ 150 bytes).

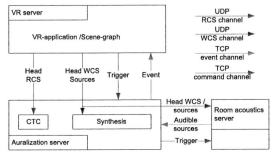

Figure 6.9:
Communication scheme which is split into several channels to optimally adapt to the internal data scheduling of the VR-application.

6.3   Performance

Central aspects of coupled real-time systems are the latency and the update rate for the communication. In order to get an objective criterion for the required update rates, it is necessary to analyze typical behavior inside *CAVE-like* environments with special regard to head movement types and magnitude of position or velocity changes. In a second step an overview of the complete system's performance is given.

6.3.1   Real-Time Requirements

In general, user movements in *CAVE-like* environments can be classified into three categories [LVJ03]. One category is defined by the movement behavior of the user inspecting a fixed object by moving up and down and from one side to the other in order to accumulate information about its structural properties. A second category can be seen in the movements when the user is standing at one spot and uses head or body rotations to view different display surfaces of the *CAVE-like* environment. The third category for head movements can be observed when the user is doing both, walking and looking around in the *CAVE-like* environment. The typical applications used in this context can be classified mainly as instances of the last two categories, although the exact user movement profiles can differ individually. Theoretical and empirical discussions about typical head movement in virtual environments are still a subject of research, e.g. see [AB95, CHV02, LVJ03] or [WO95].

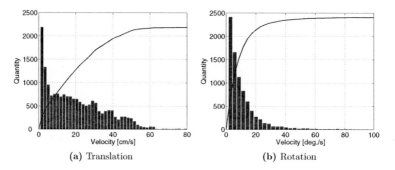

(a) Translation   (b) Rotation

Figure 6.10: Histogram of translational (v_t) and rotational (v_r) velocities of movements of a user acting in a *CAVE-like* environment. The the overly curve depicts the cumulative percentage of the measurements. In (b), the upper bound is limited to 100 deg./s for better readability.

In a field study, tracking data of the users' head movements are recorded while interacting in the virtual environment. The magnitude of the velocity of head rotation and translation is calculated with these data in order to determine the requirements for the filter calculation. Figure 6.10(a) shows a histogram of the evaluated data for the translational velocity. Following from the deviation of the data, the mean translational velocity is at 15.4 cm/s, with a standard deviation of 15.8 cm/s and the data median at 10.2 cm/s, compare Table 6.3.1.

In addition to the translational behavior, Figure 6.10(b) shows the rotational profile for head movements of a user. Peak angular velocities can be up to 140 deg./s although these are very seldom. The mean for rotational movement is at 8.6 deg./s with a standard deviation of 11.1 degrees/s and a data median at 5.2 deg./s, compare Table 6.3.1. Data sets provided as standard material for research on system latency, e.g. by [AB94] or [LVJ03], show comparable results.

	Translational velocity (v_t)	Rotational velocity (v_r)
Mean $(\bar{v})$	15.486 cm/s	8.686 deg./s
Median $(\tilde{v})$	10.236 cm/s	5.239 deg./s
Max (v_{max})	84.271 cm/s	141.458 deg./s
St. dev. (σ)	15.843 cm/s	11.174 deg./s

Table 6.3: Statistics for the measurements of the translational and rotational velocities.

6.3.2 Performance of the complete System

Several aspects have to be taken into account to give an overview of the performance of the complete system, the performance of each subsystem, the organization of parallel processing, the network transport, but also the velocity of sources and finally the user. For considerations about audio to video synchronization two issues have to be taken into account..

The first issue is the latency of the system concerning starting, pausing, stopping and altering of attributes of virtual sound sources. This issue is important for the matching of a suddenly appearing sound to a specific object or event. [VdPK00] have shown that it is an advantage if a sound that indicates a specific situation (e.g. a sound that is emitted from a hammer that hits a steel plate) is optimally presented 35 ms after the situation has been visually perceived by the user.

The worst case transmission time measured in Section 6.2.4 for the TCP command transmission is 1.2 ms. The eventually necessary update of the binaural filter consumes additional time of ≈ 1.5 ms (free-field). The output block length of the convolution is 256 taps as well as the chosen buffer size of the sound output device, resulting in a time between two buffer switches of 5.8 ms at 44.1 kHz sampling rate.

This results in a total time of 8.5 ms before the sound is transmitted over the loud-speakers for a command event such as starting a sound. This is by far fast enough for the generation of congruent audio/video events.

The other synchronization issue to deal with is the latency for updates that are necessary if the user is moving. The latency discussion as far as this issue is concerned has to consider the head-tracking technology that is used in the system. [BSM+04] have stated that an update lag of 70 ms at maximum remains unnoticed for the user.

With respect to the timing, the optical tracking system is capable of delivering spatial updates of the position and orientation of the user's head, and an additional interaction device to the VR application in 18.91 ms. This figure is a direct result from the sum of the time needed for the visual recognition of two tracking targets, as well as the transmission time for the measured data over a network link. For applications that require a minimum of latency time and do not need wireless tracking, the usage of an electromagnetic tracking system can reduce the latency to $\approx$ 5 ms.

The latency of the audio system is the time elapsed between the incoming of a new position and orientation for either a source or the listener, and the point in time the output signal is generated with the updated filter functions. The calculation of a new CTC filter set (1024 taps) takes 3.5 ms and 1.5 ms to process a new binaural filter for a single sound source in the free field. Finally, the worst case buffer swap latency of the sound output device adds another 5.8 ms. This results in a total filter processing time of 10.8 ms. Table 6.4 shows the costs for an end-to-end trip starting with the tracking and ending with the transmission of the sound over the loudspeakers. Values concerning the latency produced by the internal processing of the VR application (transformation, serialization, deserialization) are taken from [AKL05]. The table shows worst case measurements which still meet the 70 ms update criteria.

The values given in Table 6.4 are related to the free-field simulation. If the room acoustics are considered the complete performance is a combination of the room acoustical simulation which is processed on an external subsystem and separate hardware and the filter calculation which is processed by the audio server.

The update rate of the room acoustical simulation can be rather low for translational movements as the overall sound impression does not change dramatically in the immediate vicinity (see [Wit04, WD07] for further information). To provide an example, the a room acoustical simulation of a concert hall is assessed. The threshold for triggering a recalculation of the raw room impulse response is set to 25 cm, which

Cost Item		Time
Tracking (60Hz)		18.20 ms
UDP transport	+	0.701 ms
Transformation	+	0.10 ms
(De)serialization	+	0.22 ms
UDP spatial update	+	0.70 ms
Filter processing	+	10.80 ms
Sum	=	30.72 ms

Table 6.4: Total costs for a end-to-end trip of a spatial update from a tracking sample to the output of the loudspeakers.

is typically half a seat row's distance. With respect to the translational movement profile of a user, a recalculation has to be done approximately every 750 ms to catch about 70% of the movements (see Section 6.3.1 *Real-Time Requirements*). If the system aims at calculating correct image sources for about 90% of the movements, this will have to be done every 550 ms. A raw impulse response contains the raw data of the images, their amplitude and delay, but not their direction in listener's coordinates. The slowly updated dataset represents, thus, the room-related cloud of image sources. The transformation into 3D listener's coordinates and the convolution will be updated much faster, certainly, in order to allow a direct and smooth responsiveness and is directly processed on the audio server.

The processing time of the binaural impulse response depends on the simulated room. A concrete example of a room acoustical simulation, namely of the concert hall of Aachen's *Eurogress* convention center is used for the estimation of the filter processing time. Figure 6.11 shows the polygon model which has been used for the simulation and also the calculation times required for the audibility test. More detailed information about this model and the reduction to the acoustically relevant structures can be found in [SL06, LSVA07, SDV07]

With the assigned time slot of 750 ms for the simulation process, real-time capability for this room acoustical simulation is reached for about $\approx 310,000$ image sources (order 3) to be tested during runtime. 111 audible image sources are found in the first part of the impulse response of 6,000 samples length corresponding to 136 ms. In this case one source is placed on the stage, and the listener is located in the middle of the room. The complete filter processing (excluding the audibility test) is done in 6.95 ms according to Table 6.1. The complete update time of the system according

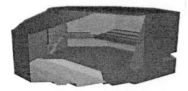

IS Order	Number of IS		Audibility test
	all	audible	
1	75	9	0.15 ms
2	4,827	32	10.46 ms
3	309,445	111	710.07 ms

Figure 6.11: Sliced polygon model of the concert hall of Aachen's *Eurogress* convention center and of the measurement results of the IS audibility test.

to Table 6.4 is summed up to 36.17 ms.

It should be noted, that the filter processing has different entry points. The rotation of the listener or a source does not cause a recalculation of the audible sources, only the binaural filter has to be processed.

Furthermore, the filter will be updated at any time the source or the head has moved more than 2 cm or has turned more than 1 degree, respectively. The contribution of each reflection is calculated on the current list of audible image sources, updated in their positions. The resulting filter only contains a few wrong reflections which will be removed after the next audibility test when the list of audible sources is refreshed. Thus, the specular reflections at the first part of the impulse response become audible with the correct spatial representation already after ≈ 36 ms (including tracking, UDP transport, CTC filter generation, binaural filter generation, and audio buffer swap).

The maximum velocity for indirect navigations has to be limited in order to avoid artifacts or distortions in the acoustical rendering and perception. However, during the indirect movement, users do not tend to move their head and the overall sensation reduces the capability to evaluate the correctness of the simulation. Once the users stop, it takes about 750 ms as depicted above to calculate the right filters for the current user position. The experience which have been made so far show that a limitation of the maximum velocity for indirect navigation to 100 cm/s shows good results and user acceptance.

Chapter 7

Validation

Up to now the technical performance has been tested and an evaluation has been made to examine whether the crosstalk compensation works in a not ideal anechoic environment. For a more complete analysis listening tests have to be accomplished to get the subjective localization performance in the VR environment. In this chapter the complete system is tested to validate the general applicability as spatial audio rendering system for Virtual Reality. All listening tests are performed in the dynamical mode, which means that the user has the possibility to turn his head to improve his perception. The tests were carried out with 15 subjects who were mostly inexperienced regarding localization tests. All source positions were presented randomized to avoid learning effects. The whole test is separated into two parts. Only auditory stimuli are presented in the first test, while a combination of visual and auditory stimuli is evaluated in the second test. The complete test is driven by the VR toolkit ViSTA, more precisely the control of the application flow, the preparation of the visual feedback as well as the data logging of all spatial information.

7.1 Auditory stimuli

Presenting only auditory stimuli under real operating conditions provides information about the system's performance in this special environment where the reflecting walls cause a depreciation of the resulting localization performance. For comparison, a detailed study on the localization performance using headphones for reproduction can be found at [WAKW93] and [Beg91]. This experiment focuses on the results for localization in the horizontal plane. In general, the localization of elevation is not very precise without any visual or acoustical anchor stimulus. In a multi-modal virtual environment it is also more important that the user is assisted by the acoustical component in areas which are not in the field of view. Hence the front/back confusion is an important factor of the spatial audio rendering system's performance. For this reason the localization is tested with stimuli which are all presented from behind the user [SLA06]. The virtual sources are placed as follows: -120deg., -150deg., 180deg., 150deg. and 120deg. azimuth, each with an elevation of -30deg., 0deg. and 30deg.. The test candidate is allowed to turn his head slightly but not to perform a complete turn. The judgment was recorded by the experimenter. The subjects are asked to determine the direction of the virtual source relative to the front direction. A marker at the frontal display panel of the *CAVE-like* environment is used as anchor. The judgments are reported to the experimenter in analogy to a clock dial, e.g. six o'clock for a source from behind and three o'clock for a source on the right hand side of the listener.

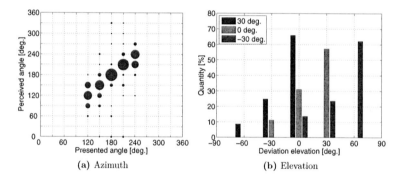

(a) Azimuth   (b) Elevation

Figure 7.1: Localization results of virtual sources placed behind the user (a) and errors concerning the perceived elevation angle sorted by the elevation of the virtual sources.

The scatter plot (Figure 7.1(a)) depicts the perceived azimuth angles in relation to the angles where the virtual sources are placed. This plot includes all stimuli regardless of the degree of elevation. The diameter of a circle represents the number of trials that have caused the specific mapping of real and perceived positions. All trials which have been perceived correctly are located on the diagonal of the plot.

It can be seen that the amount of front/back confusions that often appears with a static binaural synthesis is not very high. The fact that the user was allowed to move and turn the head slightly causes a variation of the ITD and therefore an explicit placing of the virtual source in the horizontal plane. The system is able to provide adequate cues for a good azimuth localization and a stable front/back determination.

In the vertical direction the stimuli are presented at -30 deg., 0 deg. and 30 deg. elevation. The results are widespread, as expected. Figure 7.1(b)) shows that above all the range of mismatch is significantly higher for sources placed below (negative elevation) the horizontal plane. The test of virtual sources with a positive elevation angle results in a more congruent perception. As mentioned before the speakers are placed above the screens. The reflections in this environment may add elevation cues from the reproduction loudspeakers to the signal at the ears of the listener.

7.2 Visual and Auditory Stimuli

Figure 7.2: The listening test environment. White = visual stimulus, red = free to move control stimulus.

A presentation of auditory stimuli together with visual stimuli is the default setup that the system is designed for and hence, the most important part of the listening tests [LAS05]. It is tested whether the auditory perception matches with the visual image. When the auditory stimulus appears and the related image is in the user's field of view, a large deviation of sound and image is detected directly, a small deviation may be masked. Figure 7.2 shows the visual feedback for the user. The visual stimulus is a white ball. The control feedback, a red ball, can be adjusted by the user using a flight-stick.

In this test the user is asked to detect the deviation of the visual and the auditory stimuli. The user indicates the perceived position by moving the control feedback (red ball) to the position of the virtual source. The user is free to walk around and turn his head. The position of the control feedback and of the visual stimulus itself is stored in a logging file after confirming the choice to determine the deviation. It is not known to the user that the system always placed the sound exactly at the visual stimuli. The user is told that deviation is added by the system at random to analyze whether he or her is able to detect it.

The stimuli are presented at the following positions: -67 deg., -22 deg., 0 deg., 45 deg., and 90 deg. azimuth, each with an elevation of -30 deg., -15 deg., 0 deg., 15 deg., and 30 deg.. The distance was fixed at 1.2 m to the center of the *CAVE-like* environment.

In a dynamic environment where the user is able to turn his head, a "usual" listening test setup with a constant distribution in space is not useful. The listener is able to turn towards the source and the position of the source is thus not relevant anymore. The reason for choosing these special positions for the test is related to the speaker setup for the crosstalk cancellation. As mentioned above, certain loudspeakers are active depending on the user's position.

There are three possible CTC modes, one uses speakers spanning an angle of 90 degree; the other one spans an angle of 180 degree related to the center of the *CAVE-like* environment. The third one uses both configurations with a position-dependent fading between them, as described in Chapter 4.2. Stimuli are placed at 45 deg. to test the 90 deg.-configuration, stimuli are placed at 0 deg. and 90 deg. to test the 180 deg.-configuration. Stimuli at 67 deg. and 22 deg. are used to test the performance in the fading area between the two configurations.

Figure 7.3 shows the results for all tests with a stimuli presentation each in a certain elevation. The results for the sources placed at ±15deg. elevation are not presented here as they show very similar results. Approximately the same diffusion in the azimuth localization occurs for all different elevation angles of the virtual sources.

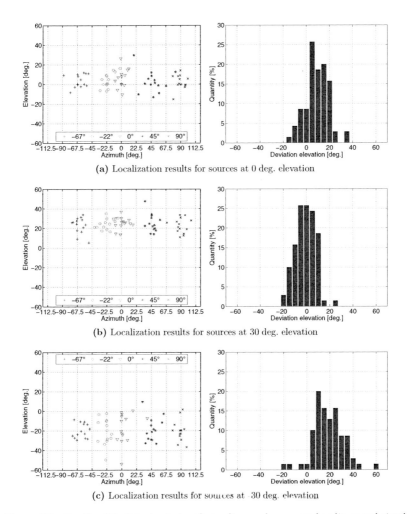

(a) Localization results for sources at 0 deg. elevation

(b) Localization results for sources at 30 deg. elevation

(c) Localization results for sources at 30 deg. elevation

Figure 7.3: Results of the listening test with simultaneously presented auditory and visual stimuli for each tested elevation angle.

The deviation between the presented azimuth angle and the estimated angle regardless of the source elevation is plotted in Figure 7.4(a) and grouped in 2deg. steps. The median value of the azimuth angles is 0.044deg. with a standard deviation of 9.61deg.. The same is shown for the elevation in Figure 7.4(b). The over-all median value for all presented elevation angles is still good (1.27deg.) but Table 7.1 shows that the median value and the standard deviation differ tremendously depending on the presented elevation. As seen in the results of the test without visual feedback, estimations for the elevation angle for sources placed at a lower elevation angle are often too high. But the perception of distributed sources in elevation is better for a combined visual and auditory presentation than without the visual feedback. However, the perceptual evaluation of the localization performance by listening tests shows that the spatial rendering in combination with the crosstalk compensated reproduction over loudspeaker leads to reliable results. The localization concerning the azimuth angles is very good despite the reflective environment.

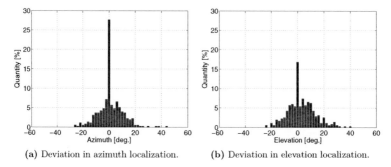

(a) Deviation in azimuth localization.   (b) Deviation in elevation localization.

Figure 7.4: Histogram of deviation for the combined visual and auditory stimuli experiment.

Furthermore, all participants of the listening test were asked to describe the impression as a whole. Only very few subjects reported that they heard any filter switching noises or a coloration of the sound and if so, it was noticed only during fast movements. This substantiates the assumption which has been made in Chapter 5. The variation of the channel separation between the filters is below of a noticeable level.

Presented	Perceived angle	
angle	Median	Std. dev.
−30.deg.	−20.17.deg.	11.78.deg.
−15.deg.	−9.24.deg.	9.28.deg.
0.deg.	0.99.deg.	8.34.deg.
15.deg.	14.89.deg.	7.27.deg.
30.deg.	24.45.deg.	7.24.deg.

Table 7.1: Median values and standard deviation of the perceived elevation angles quoted separate for different virtual source elevations.

The validation of the complete system concerning all different aspects of perception is by far not completed. The listening experiments which are described in this chapter are carried out to verify the statement that the binaural synthesis and reproduction is suitable for VR systems in general. This is also the fact why the main focus of the validation is put on the localization performance. The influence of the visual representation on the perception of the spatial audio representation has to be investigated more into detail in future.

A topic of current investigation is the perceptual and attentional interaction of different senses. It is studied whether a stimulation by, for instance, short sound of a certain direction can "shift" the spatial focus of attention of the visual sense and vice versa. This is indicated by a faster reaction on stimuli of the same direction. The study of these cross-modal effects is also an interesting issue to validate the audio video coupling of the complete system.

Furthermore, one important study is the comparison of the differences concerning the localization of virtual sources and real sources at the same positions and in the same environment. Thus, the perceived positions of virtual sources have to be compared with the perceived position of real sources and not with the position which is defined by the chosen HRTF. This means that the localization blur of real sources and virtual sources has to be compared.

The validation is carried out only for the synthesis of the free-field sources so far. In a next step the room-acoustical representation has to be considered during the listening tests. Furthermore, the interface to the room-acoustical simulation enables very interesting studies of the influence of the visual representation on the acoustical perception of a room and vice versa.

Chapter 8

Summary

In this thesis a spatial audio rendering and reproduction system has been realized, integrated into an existing Virtual Reality system and evaluated regarding objective, and subjective criteria. A multi-modal VR environment with congruent visual and auditory representation requires a precise spatial placement of the virtual sources and at least a plausible representation of all other acoustical properties, such as radiation of sources and the influence of the virtual room.

The first chapters describe the required technology with special focus on real-time applicability. This comprises of:

- The simulation of the source and its radiation characteristics (Chapter 2)

- The transmission to the ears of the virtual listener, accomplished by the binaural synthesis (Chapter 3)

- The reproduction at the ears of the real listener realized by the dynamic crosstalk cancellation (Chapter 4)

Especially in an interactive environment where the user is able to displace the virtual source or where the source itself is moving, the consideration of radiation and distance provide additional cues for enhancing the plausibility of the complete simulation.

Several natural instruments have been recorded to generate a directivity database for the simulation. In contrast to electro-acoustical sources which can be measured sequentially from each direction, the signal of a natural instrument has to be recorded simultaneously from all directions to obtain comparable results with regard to a congruent excitation.

The influence of averaging the directivity information extracted from different single tones of a scale has been analyzed and compared to the directivity extracted from a short phrase with a representative pitch scale. Furthermore, the influence of the pitch has been evaluated in order to specify whether the directivity is independent of the excitation. It was found that this is a valid assumption for rather directional radiating instruments. But non-directional radiating instruments such as a violin show a directivity which is affected by the tone which is being played. However, a directivity being valid for the complete pitch-range of a specific instrument is required in case of separating the source content and the radiation. In this case, the frequency characteristics of the recorded instrument at the reference direction, together with the relative or normalized directivity, leads to the complete specification of the instrument for any direction.

If the simulation of a source is not of sufficient quality by using an averaged directivity it is possible to use multi-track recordings with one track for each direction. The advantage is that the excitation and the radiation characteristics are not separated. The disadvantage, however, is the high amount of processing power being required for the convolution of the additional channels. The realized audio server has the capability to support both methods depending on the focus of application.

The ability to assign a direction to a sound event is based on the binaural perception and is caused by the different temporal and frequency-dependent cues of the signals at the ears. It is possible to generate a binaural signal which contains all spatial cues being required for a correct localization of the virtual source by applying the head-related transfer function of the direction being desired. To realize a room-related virtual source, the HRTF has to be adapted when the listener turns or moves his head. Apart from the generation of room-related sources, the main advantage of a dynamic synthesis is an almost complete elimination of front-back confusion as it often occurs while using static binaural synthesis with non-individualized HRTFs.

The ability of humans to recognize a source position near to the head is based on distance-dependent differences of the HRTFs which have to be taken into account for a plausible simulation of near-to-head sources. For the synthesis database, the HRTFs of the ITA artificial head were measured at distances of 0.2 m, 0.3 m, 0.4 m, 0.5 m, 0.75 m, 1.0 m, 1.5 m, and 2.0 m. The spatial resolution is chosen to 1 degree for the azimuth angle and 5 degree for the elevation angle.

A special focus is put on the context of dynamic aspects regarding position and orientation of the listener and the source. Besides the advantage of the dynamic synthesis in view of better localization results, the technical drawback is the filter change being required in order to achieve a room-related virtual source. Whenever a source is moving or the head moves relative to a source, the transfer functions from the source to the ears change. In the case of binaural synthesis, this means that the HRTFs used for the filtering have to be changed. But every filter change also causes artifacts in the resulting audio signal. Several methods have been examined to achieve an exchange of HRTFs without audible artifacts.

An exact spatial impression is possible if the generated binaural signal is reproduced exactly at the ears of the listener. If loudspeakers have to be used for the reproduction instead of headphones, the interfering crosstalk has to be canceled. To realize of a crosstalk compensation which is not only valid at one point, it is necessary to update the CTC filter depending on the listener's position. The filter calculation at run time of the program is based on a database which contains HRTFs measured in a spatial resolution of 1 degree for azimuth and elevation. The stereo CTC setup was extended to four loudspeakers to provide a complete 360 degree turn of the listener. Depending on the head position the best two-loudspeaker configuration is chosen to accomplish the CTC.

The crosstalk cancellation system has been evaluated by means of objective and subjective criteria. The measurements of the channel separation which can be achieved using the dynamic system show very good results on an average of 20 to 25 dB at a wide frequency range.

A *CAVE-like* display system with reflecting video projection screens is not an acoustically ideal (anechoic) environment, but it is the common field of application of this audio system. The applicability of the crosstalk cancellation in such an environment was evaluated by carrying out listening tests. These tests show that the dynamic crosstalk cancellation, in combination with a dynamic synthesis achieves good results also in not ideally anechoic environments. This system acts as a virtual headphone providing the channel separation without the need to wear physical headphones.

Finally, the described technology is used for the implementation of an audio rendering and reproduction system. The cooperation of all subsystems which are shown in Figure 8.1 enables the generation of a complex and multi-modal virtual scene. The main application provides all information, the temporal flow of the presented content, and the associated data. This application is running on the same platform which is responsible for the visual rendering and is also responsible for the distribution of the data to the subsystems.

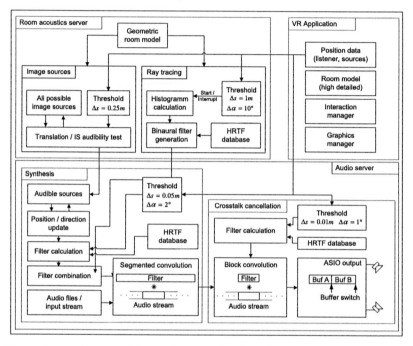

Figure 8.1: Architecture of the complete spatial audio rendering and reproduction system including the external components.

As mentioned above, a complete description of a scene is often not possible without the room acoustics being taken into account. The calculation of the information being required to model the room acoustics is calculated on a dedicated subsystem, the room acoustics server which is used here as a network service. The binaural room impulse response is calculated on the audio server in order to be able to react as fast as possible on any movement of the user or of a virtual source. A fast low latency convolution engine ensures that impulse responses regardless of their complete length will be considered by the filtering of the mono audio material after 5.8 ms (256 samples). Optimizations concerning modern processor extensions enable the rendering of 10 sources with filters of 3 s (132000 taps) length or 15 sources with filters of 2 s length, for example.

The time of a complete update cycle including the tracking latency, binaural filter processing, the calculation of the CTC filters and the latency added by the sound output device accumulates to ≈ 36 ms. The control commands (e.g. start/stop) will be considered in the audio server after 1.5 ms in a worst case scenario. The changes are served with the next output block (5.8 ms) which guarantees a tight audio video synchronism.

The validation of the complete system shows that placing virtual sources at an arbitrary position in the three-dimensional space is possible. The reproduction of sources in particular at a negative elevation angle (below the listener) is still not as precise as desired whereas the localization results of the azimuth angle are very good.

To draw a conclusion it can be stated that this system represents an efficient and very robust binaural rendering and reproduction by means of loudspeakers in the full space around the user. It is suited to enhance VR systems such as a *CAVE-like* environment with spatial audio to combine visual and acoustical stimuli in an excellent way.

Outlook

The realized spatial audio rendering and reproduction system is still open for further enhancements and extensions.

The prediction of the listeners position would be useful to overcome the 60 Hz update limitation of the optical tracking and thus, enhance the reproduction concerning the crosstalk cancellation for very fast user movements. A further improvement of the CTC is the cancellation of the reflections caused by the video projection screens. A first study has already been made [Fro05] but has to be integrated and tested whether the precision of the position determination of the head and the image sources which represent the mirrored reproduction loudspeakers leads to reliable results.

The binaural synthesis shows already good performance but the representation of low elevated sources could be further improved. Taking the directional bands [RB68, Bla97] into account could be an appropriate method to improve the perception at the desired elevation angle.

A dynamical clustering of virtual sources would be an extension which is feasible to auralize more sources in a complex scene. An object may consist of several virtual sources of different content, directivity, and position in the near-field of the listener. Beyond a certain distance the complete object can be rendered by mixing the content of several sub-sources and use only the averaged directivity and the objects center position. In the same context, the optimization of the segmented convolution must be mentioned which is also important to generate more virtual sources, especially in conjunction with the room acoustical simulation of large environments with long reverberation times.

Finally, if the room acoustical simulation and the VR application support the exchange of material data or even parts of the geometry the audio server would be able to adapt on the changed parameters. The filter generation is already prepared for this extension. It would be possible to auralize the change of the sound field instantly, e.g. by selecting different material parameters of a wall while being present in the virtual scene. This would be of special interest for architectural design and research in the field of multi-modal perception.

Kurzfassung

Einleitung

Die Erzeugung und Nutzung künstlicher virtueller Umgebungen gewinnt immer mehr an Bedeutung und wird vor allem in Bereichen wie dem Produktdesign, der Prototypen-Evaluierung und in der Forschung zur Veranschaulichung komplexer Datensätze eingesetzt. In der Vergangenheit lag der Schwerpunkt auf der rein visuellen Darstellung, um beliebige Geometrien dreidimensional anzuzeigen (stereoskopische Darstellung). Da sich die Wahrnehmung jedoch aus einer Vielzahl verschiedener Sinneseindrücke zusammensetzt, ist es wünschenswert, auch die Repräsentation der virtuellen Szene multi-modal zu gestalten.

Im Rahmen dieser Arbeit werden zunächst Techniken beschrieben und evaluiert, mit denen realistische oder zumindest plausible akustische Szenen mit räumlich verteilten Schallquellen in Echtzeit realisiert werden können. Des Weiteren wird die Realisierung eines Softwaresystems beschrieben, das die zuvor bereitgestellten Techniken nutzt, um die visuelle Darstellung um die akustische Komponente zu erweitern.

Um eine möglichst authentische Simulation zu gewährleisten, ist es erforderlich, alle beteiligten Komponenten, die das Schallfeld beeinflussen, mit einer höchstmöglichen Genauigkeit nachzubilden. Dazu gehört die Beschreibung der Quelle mit ihrem charakteristischen winkel-, abstands- und zeitabhängigen Abstrahlverhalten, die Schallausbreitung in der virtuellen Szene (Raumakustik), die gehörbezogene Berücksichtigung aller Schallfeldanteile (binaurale Synthese) und die exakte Reproduktion dieses künstlichen Schallfeldes an den Ohren des Benutzers (Übersprechkompensation).

Abbildung 1 zeigt die Aufteilung der kompletten Signalkette von der Entstehung des Schalls bis zu den Ohren des Zuhörers in einzelne Komponenten. Diese werden im Folgenden näher erläutert und es werden Verfahren aufgezeigt, die letztendlich

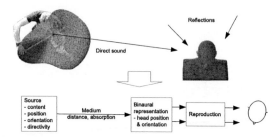

Abbildung 1: Aufteilung der natürlichen Wahrnehmung einer Quelle in die verschiedenen Bereiche, die bei der Simulation berücksichtigt werden müssen.

in die Implementierung eines komplexen Audio Rendering Systems für multimodale VR-Systeme einfließen.

Das vorgestellte System verwendet die Binauraltechnik (binaural = „beidohrig"), mit dem Ziel, an den Ohren des Benutzers das Schallsignal zu reproduzieren, das auch im Original-Umfeld dort herrschen würde.

Virtuelle Quellen

Die Abstrahlcharakteristik der Quelle ist von entscheidender Bedeutung für eine realistische Simulation. Dies gilt besonders im Zusammenhang mit dynamischen Szenen und natürlichen Quellen wie z.B. Musikinstrumenten. Wenn der Benutzer in der Lage ist, sich um die Quelle herum zu bewegen oder die Quelle sich selbst bewegt oder rotiert, wirkt sich die Vernachlässigung der Abstrahlcharakteristik (Directivity) sofort negativ aus.

Im Rahmen dieser Arbeit wurde die Abstrahlcharakteristik verschiedener Instrumente bestimmt und in einer Datenbank für die weitere Verwendung bei der Quellsynthese zur Verfügung gestellt. Für die folgenden Instrumente wurden Datensätze erstellt: Geige, Bratsche, Cello, Trompete, Piccolo-Trompete, Flügelhorn.

Im Gegensatz zu einer elektroakustischen Quelle, bei der eine Messung der winkelab-hängigen Übertragungsfunktion sequenziell durchgeführt werden kann, ist dies bei einem natürlichen Instrument nicht möglich. Aufgrund der nicht eindeutig reproduzierbaren Tonerzeugung müssen in diesem Fall die Signale aus „allen" Richtungen synchron aufgezeichnet werden.

Zu diesem Zweck wurden die Instrumente in einem reflexionsarmen Vollraum mit 24 Mikrofonen parallel aufgezeichnet. Die Mikrofone waren dabei auf der Oberfläche einer Kugel (Radius 1,5 m) verteilt. Ziel dieser Messung war nicht vorrangig eine

Analyse des Abstrahlverhaltens, sondern vielmehr die Erstellung eines Datensatzes, der den Qualitätsanforderungen einer Auralisation genügt. Als Grundmaterial für die Erstellung einer Instrumenten-Directivity wurden zwei verschiedene Datensätze aufgenommen; zum einen alle Einzeltöne im Tonumfang des Instrumentes; zum anderen eine kurze Phrase eines Musikstücks mit repräsentativem Tonumfang. Abbildung 2 zeigt die Directivity einer Trompete und einer Geige als dreidimensionale Balloon-Darstellung jeweils für Frequenzen von 400 Hz, 1,25 kHz und 4 kHz.

Die Auswertung und der Vergleich der Datensätze zeigt, dass die Directivity für stark richtende Instrumente (z.B. Trompete) sehr gut aus der Überlagerung der Beiträge aller Einzeltöne erzeugt werden kann. Ungerichtet abstrahlende Instrumente (z.B. Geige) zeigen jedoch sehr starke Unterschiede im Richtverhalten in Abhängigkeit des gespielten Tons. Bei der Analyse einer kompletten Phrase werden die Anteile aller Töne natürlich gemittelt. Dies ist ein geeignetes Verfahren, eine allgemeingültige Directivity (unabhängig von der Anregung) zu generieren. Eine allgemeingültige Directivity ist im Kontext der Simulation und Auralisation außerordentlich wichtig, da das Quellmaterial im Normalfall aus einer Audioaufnahme besteht und nicht ohne weiteres auf die reine Anregung (Tonhöhe) zugegriffen werden kann.

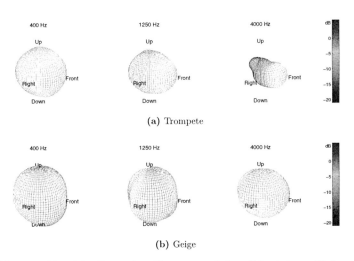

(a) Trompete

(b) Geige

Abbildung 2: Directivity Daten einer Trompete und einer Geige bei verschiedenen Frequenzen. Die örtliche Auflösung ist auf Winkelschritte von 5° interpoliert.

Eine Möglichkeit diese Einschränkung zu umgehen, ist die Verwendung von Mehrkanalaufnahmen bei der Synthese von Quellen. In diesem Fall ist der Quell-Inhalt und die Abstrahlung nicht getrennt, sondern kombiniert. Durch die Belegung mehrerer Kanäle mit nur einem Instrument wird allerdings die Gesamtzahl der virtuellen Quellen herabgesetzt, die in Echtzeit wiedergegeben werden können. In der später beschriebenen Realisierung des Audio-Servers sind beide Methoden möglich, sowie eine Kombination aus beiden, was die größtmögliche Flexibilität bezüglich der Skalierung „Echtzeit versus Genauigkeit" bedeutet, die je nach Anwendung gewählt werden kann.

Binaurale Synthese

Das menschliche Gehör ist in der Lage, Schallereignissen eine Richtung zuzuweisen. Dies gründet sich auf der Tatsache, dass das Schallsignal binaural, also mit zwei Ohren ausgewertet wird. Als Gehör wird im Folgenden der komplette Hörsinn bezeichnet mit allen dazugehörigen Schritten der zentralen Verarbeitung durch das Gehirn. Die Richtungswahrnehmung wird im Wesentlichen durch zwei Faktoren bestimmt. Die interaurale Zeitdifferenz (ITD = Interaural Time Difference) beschreibt den Laufzeitunterschied des Schalls zischen linkem und rechtem Ohr. Der interaurale Lautstärkeunterschied (ILD = Interaural Level Difference) wird durch den abschattenden Einfluss des Kopfes hervorgerufen.

Ein binaurales (zweikanaliges, kopfbezogenes) Signal repräsentiert das Schallfeld am Eingang der beiden Ohrkanäle und beinhaltet die genannten zeitlichen und spektralen Merkmale, die eine Lokalisation der Quelle im Raum ermöglichen. Ein solches Schallsignal kann z.B. mit Sondenmikrofonen in den Ohren einer realen Person oder mit einem Kunstkopf aufgenommen werden. Die Übertragungsfunktion von der Quelle zu den Ohren wird als Außenohrübertragungsfunktion bezeichnet (HRTF = Head Related Tranfer Function im Frequenzbereich bzw. HRIR = Head Related Impulse Response im Zeitbereich).

Als Fernfeld bezeichnet man den Bereich, in dem die ILD (der relative Pegelunterschied zwischen den Ohren) entfernungsunabhängig ist und damit an beiden Ohren der Pegel gleichmäßig nach dem 1/r-Gesetz abnimmt. Im Nahfeld (ca. < 1,5 m) gilt dieser Zusammenhang nicht mehr, da insbesondere die Abschattung für kopfnahe Quellen größer wird. Es werden also andere HRTFs als im Fernfeld benötigt, um eine realistische Simulation kopfnaher Quellen zu ermöglichen. Zu diesem Zweck wurden

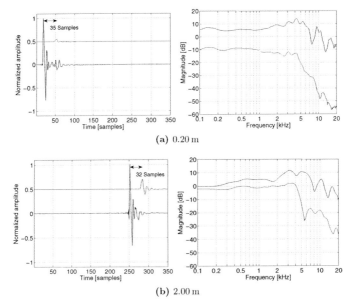

(a) 0.20 m

(b) 2.00 m

Abbildung 3: HRIRs (links) und HRTFs (rechts) im Abstand von 0,2 m und 2 m. Der Azimutwinkel beträgt 90° in der Horizontalebene. Linker und rechter Kanal der HRIR sind um 0.5 versetzt dargestellt.

HRTFs des ITA-Kunstkopfes im Abstand von 0,2 m, 0,3 m, 0,4 m, 0,5 m, 0,75 m, 1 m, 1,5 m und 2 m jeweils mit einer Winkelauflösung von 5° für die Elevation (vertikal) und 1° für den Azimut (horizontal) gemessen. Abbildung 3 zeigt exemplarisch die HRTFs bzw. HRIRs für einen Abstand von 0,2 m und 2 m bei einem Azimutwinkel von 90°.

Als binaurale Synthese bezeichnet man die Faltung eines beliebigen Audiosignals (mono) mit einer HRIR. Das Ergebnis dieser Faltung ist dann ein binaurales Signal, das alle Richtungsinformationen zur Ortung der Schallquelle in sich trägt. Bei einer statischen Synthese, beispielsweise über Kopfhörer, dreht sich das erzeugte Schallfeld, wie bei einer einfachen Kunstkopfaufnahme, mit der Bewegung des Kopfes. Für die Erzeugung akustischer Szenen, die zu einem visuellen Eindruck passen sollen, ist es jedoch erforderlich, an der Szene fixierte, ortsfeste Quellen zu realisieren. Dies ist mit der dynamischen nachgeführten Synthese möglich. Durch die Einbeziehung der aktuellen Kopfposition wird die entsprechende HRTF nach Bestimmung der Relativrichtung des Kopfes zur virtuellen Quelle geladen und das Signal mit dieser gefiltert. Der große Vorteil dieser dynamischen Synthese liegt in der Reduzierung der sonst häufig beobachteten Vertauschung der empfundenen Schalleinfallsrichtung von vorne und hinten. Dies wird durch die nicht eindeutig auswertbare Zeitdifferenz des Signals am linken und rechten Ohr hervorgerufen (siehe Abb. 4). Ist die Quelle jedoch ortsfest, kann der Hörer kleine Peilbewegungen ausführen und dadurch die Richtung eindeutig bestimmen.

Abbildung 4: Veränderung der interauralen Zeitdifferez in Abhängigkeit von der Kopforientierung bei raumfesten Quellen.

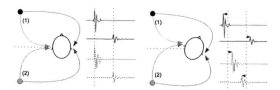

Übersprechkompensation

Der Schwerpunkt dieser Arbeit ist die Reproduktionstechnik für binaurale Signale. In einer VR Umgebung, bei der sehr großer Aufwand betrieben wurde, um dem Benutzer das Tragen sperriger und schwerer Hardware (z.B. HMDs, Head Mounted Displays) zu ersparen, ist die Verwendung von Kopfhörern für die Audiowiedergabe nicht erwünscht. Die fehlerfreie Wiedergabe binauraler Signale ist jedoch über Lautsprecher ohne zusätzliche Signalverarbeitung nicht möglich, da die Lokalisation der Schallereignisse durch die schlechte Kanaltrennung (Übersprechen) verfälscht bzw. zunichte gemacht wird. Diese unzureichende Kanaltrennung mit dem Verfahren der Übersprechkompensation deutlich verbessert werden. Diese modifiziert das wiederzugebende Signal so, dass die Anteile aus dem Übersprechen vom rechten Lautsprecher zum linken Ohr und vom linken Lautsprecher zum rechten Ohr ausgelöscht werden. Im Idealfall gelangt nur das eigentliche Nutzsignal an das betreffende Ohr des Zuhörers.

Soll das Kompensationsfilter für einen beliebigen Ort und eine beliebige Orientierung der Person gültig sein, muss eine dynamische Nachführung der Kompensation durchgeführt werden, bei der abhängig von der Position und Orientierung ein gültiges Filter zur Verfügung gestellt wird. Die vier Übertragungswege in Abbildung 5 repräsentieren zwei Sätze HRTFs unterschiedlicher Richtung. Der Abstand des Wiedergabe-Lautsprechers zum Kopf wirkt sich in der Laufzeit und im Pegel der einzelnen HRTFs aus. Laufzeit und Pegel lassen sich jedoch im Nachhinein anpassen, so dass die Einbeziehung der Abstands-abhängigkeit im laufenden Programm erfolgen kann. In dem hier vorgestellten System kommt eine Datenbank, welche die HRTFs des ITA-Kunstkopfes in einer Auflösung von 1° für Azimut und Elevation enthält, zum Einsatz.

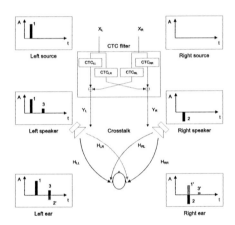

Abbildung 5: Prinzipielle Darstellung der Übersprechkompensation als iterativer Prozess. Der vom linken Lautsprecher abgestrahlte Impuls spricht mit entsprechendem Zeitversatz und durch die Abschattung des Kopfes im Pegel reduziert (frequenzabhängig) zum rechten Ohr über. Ein Kompensationsimpuls wird über den rechten Lautsprecher in der Art abgestrahlt, dass er mit dem Übersprechen zeitgleich, jedoch um 180° phasenversetzt am Ohr eintrifft. Das erneute Übersprechen des Kompenstionspulses muss nun ebenfalls kompensiert werden. Durch die Abschattung des Kopfes ist dies jedoch ein abklingender Vorgang.

Eine stabile Übersprechkompensation ist allerdings nur in dem von den Lautsprechern aufgespannten Winkel möglich. Befinden sich beide Wiedergabelautsprecher auf einer Seite des Kopfes, kann die natürliche Abschattung des Kopfes nicht mehr ausgenutzt werden. Das erneute Übersprechen des Kompensationssignals für das abgewandte Ohr wird jetzt nicht mehr gedämpft, was einen Anstieg der Filterfunktion über die Zeit zur Folge hat. Abbildung 6 (a) zeigt die Gültigkeitsbereiche für zwei unterschiedliche Lautsprecheranordnungen.

Da sich die Gültigkeitsbereiche überlappen, ist es möglich, die komplette Kopfdrehung durch 4 Lautsprecher, also die Kombination von 8 Paarungen abzudecken (siehe Abb. 6 (b)). Der Wechsel zwischen den Sektoren wird mit einer winkelabhängigen Überblendung realisiert, so dass keine Umschaltvorgänge hörbar sind.

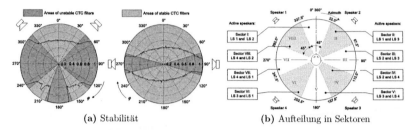

(a) Stabilität   (b) Aufteilung in Sektoren

Abbildung 6: Bereiche in denen die Übersprechkompensationsfilter stabil bzw instabil sind (a) und die sich daraus ergebende Aufteilung in Sektoren, in denen unterschiedliche Paarungen der Lautsprecher verwendet werden.

Evaluierung

Bevor die hier vorgestellte Wiedergabetechnik in ein komplettes Audio-Rendering- und Wiedergabesystem eingebettet werden kann, soll zunächst untersucht werden, ob diese den Anforderungen einer korrekten binauralen Reproduktion genügt.

Im ersten Teil wird messtechnisch die erzielbare Kanaltrennung an verschiedenen Punkten untersucht. Das komplette System wird im dynamischen Modus betrieben, d.h. alle Filter werden zur Laufzeit generiert, die Position des Kunstkopfes, der in diesem Fall die Abhörperson ersetzt, wird mittels Head-Tracker erfasst.

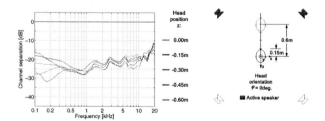

Abbildung 7: Gemessene Kanaltrennung an verschiedenen Punkten entlang der x-Achse.

In Abbildung 7 ist stellvertretend für alle anderen Messungen die erzielte Kanaltrennung an einer Reihe von Punkten entlang der x-Achse dargestellt. Die Kanaltrennung beträgt ca. 20 bis 25dB über einen weiten Frequenzbereich, was ein außerordentlich gutes Resultat darstellt und nur wenig gegenüber einer statischen, für einen festen Punkt eingemessenen Übersprechkompensation zurücksteht. Für die Messungen an anderen Positionen sind ähnliche Ergebnisse zu beobachten. In diesem Test sind allerdings alle Reflexionen ausgefenstert, d.h. es werden Freifeldbedingungen vorausgesetzt. In nicht reflexionsfreier Umgebung gelten die dargestellten Werte nur für die erste Wellenfront, die jedoch eine entscheidende Rolle für die Lokalisation spielt.

In einem zweiten Test wird die Lokalisationsleistung von Probanden in reflexionsarmer Umgebung untersucht, in der es möglich ist, gezielt reflektierende Wände hinzuzufügen. Dies ermöglicht eine direkte Vergleichbarkeit. Die Auswertung zeigt, dass die Reflexionen die Lokalisation erschweren, sofern es sich um kopfbezogene Quellen handelt. Wird aber auch die Quellsynthese dynamisch nachgeführt und dem Hörer damit die Möglichkeit zu Ortungsbewegungen gegeben, sind die Ergebnisse trotz Reflexionen sehr gut (siehe Abb. 8). Das macht die Anwendung dieser Wiedergabetechnik in VR-Systemen möglich, in denen durch die notwendigen Projektionsoberflächen auch deutliche Reflexionen auftreten

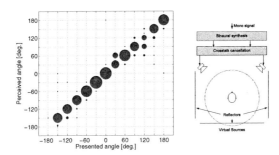

Abbildung 8: Ergebnisse des Hörversuchs und das zugehörige Setup. Das binaurale Signal wurde mittels der dynamischen (nachgefürten) binauralen Synthese generiert und über die ebenfalls dynamisch arbeitende Übersprechkompensation wiedergegeben. Der Durchmesser der Kreise ist proportional zu der Anzahl der Treffer für diese Kombination.

Interaktives VR-System

Ein Hauptziel dieser Arbeit ist die Realisierung eines Audio-Servers, der in der Lage ist, örtlich verteilte Klangquellen zu simulieren und diese korrekt wiederzugeben. Die bisher beschriebenen Techniken sollen dafür verwendet werden. Bei dem hier vorgestellten System wird also ein durchgehend binauraler Ansatz gewählt (siehe Abbildung 9).

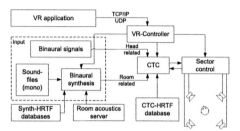

Abbildung 9: Die einzelnen Komponenten des Audio-Servers, der die dynamische binaurale Synthese und die übersprechkompensierte Wiedergabe über Lautsprecher realisiert.

Da die simulierten Szenen nicht immer Freifeldbedingungen aufweisen, sondern auch einen Raum als Umgebung beinhalten können, ist für eine plausible Auralisation die Berücksichtigung der Raumakustik erforderlich. Eine Schnittstelle zu einem weiteren Subsystem ermöglicht es, eine raumakustische Simulation mit einzubeziehen. Am Institut für Technische Akustik wird seit langer Zeit an Systemen zur Computersimulation von Raumakustik gearbeitet. Eine auf Echtzeitanforderungen optimierte Implementation wird hier als Raumakustik-Server verwendet und von dem hier beschriebenen System über Netzwerk gesteuert. Der Raumakustik-Server liefert für eine gegebene Geometrie und in Abhängigkeit der Quellpositionen eine Liste mit Spiegelschallquellen, die die frühen Reflexionen in einer Raumimpulsantwort beschreiben. Mit Hilfe dieser Spiegelschallquellen und der HRTF-Datenbank ist der Akustik-Server in der Lage, die binauralen Filter für die Quellsynthese zu berechnen. Dabei wird nicht nur die Richtung des Direktschalls, sondern auch die Richtungen aller Reflexionen korrekt einbezogen, was die Qualität und Plausibilität des Raumeindrucks im Vergleich zu einfachen Nachhallprozessoren immens erhöht. Darüber hinaus kann der späte und eher diffuse Bereich des Nachhalls aus einem Raytracing-Algorithmus gewonnen werden, der ebenfalls auf dem Raumakustik-Server gerechnet wird.

Alle Komponenten des Gesamtsystems werden von der VR-Applikation, die auch für die Visualisierung der Szene verantwortlich ist, gesteuert. Bei der Visualisierung handelt es sich um die dreidimensionale Darstellung der Objekte, die durch die stereoskopische Darstellung einen holographischen Charakter erlangen. Abbildung 10 verdeutlicht die Sicht durch das Displaysystem hindurch auf die dreidimensionale Szene. Das Displaysystem selbst besteht aus 5 Projektionsflächen (vier Seiten + Boden) auf denen jeweils mit zwei Projektoren die Bilder für das linke und rechte Auge projiziert werden. Die stereoskopischen Bilder werden am Auge wieder mit Polarisationsfiltern getrennt (passiv Stereo).

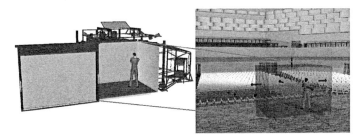

Abbildung 10: Der Blick „durch" das Display in den dreidimensional dargestellten Raum (links) und eine schematische Zeichnung des am Rechen- und Kommunikationszentrums der RWTH installierten Display-Systems.

Der Audio-Server ist in der Lage, z.B. auf eine Kopfdrehung innerhalb von ca. 36 ms (worst case) zu reagieren, wobei die Neuberechnung der Übersprechkompensation wie auch die Quellsynthese enthalten sind. Dies führt zu einem sehr stabilen Raumeindruck auch bei schnellen Bewegungen.

Da bei der Berücksichtigung der Raumakustik lange Filterimpulsantworten verwendet werden müssen, wird die binaurale Synthese durch eine segmentierte Faltung im Frequenzbereich mit geringer Latenz bewerkstelligt. Dieser bekannte Algorithmus wurde erweitert und auf die gegebene Hardware angepasst, um eine möglichst hohe Anzahl an Filterkoeffizienten zu erreichen. In der jetzigen Version können beispielsweise 10 Quellen mit einer Filterlänge von jeweils 132.000 Taps (binaural) und einer Latenz von 256 Samples (5,8 ms) bearbeitet werden.

Validierung

Abschließend wird die Fähigkeit von Probanden zur Lokalisation virtueller Quellen untersucht, die mittels des in das Display-System integrierten Audiosystems erzeugt wurden. Während der Hörversuche wurden rein akustische Stimuli als auch eine Kombination aus akustischen und optischen Stimuli dargeboten. Der Scatter-Plot in Abbildung 11 zeigt exemplarisch die Ergebnisse für virtuelle Quellen, die in der Horizontalebene unter verschiedenen Azimutwinkeln plaziert sind. Zusätzlich ist noch die

Abweichung der gehörten Elevation aufgetragen. Während den Probanden die Bestimmung der Elevation der virtuellen Quellen oft nicht gelingt und die Ergebnisse diesbezüglich stark streuen, sind die Lokalisierungseigenschaften in der Horizontalebene sehr gut.

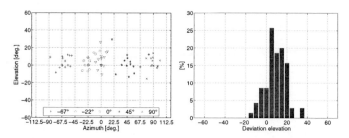

Abbildung 11: Ergebnisse des Hörversuchs mit gleichzeitig präsentierten visuellen und auditiven Stimuli für virtuelle Quellen in der Horizontalebene. Die Probanden waren aufgefordert, den örtlichen Versatz zwischen visuellem Stimulus und dem akustischen Stimulus anzugeben.

Zusammenfassung

Im Rahmen dieser Arbeit wurde ein Audio-Rendering- und Wiedergabesystem realisiert, das die freie Platzierung virtueller Quellen im Raum ermöglicht.

Beginnend mit der Simulation der Quelle inklusive des Abstrahlverhaltens wurde die komplette Übertragung zum Ohr durch die binaurale Synthese nachgebildet. Die Schnittstelle zur raumakustischen Simulation erlaubt darüber hinaus auch die Einbeziehung der Reflexionen in die akustischen Szenen.

Die Wiedergabe der synthetisierten binauralen Signale erfolgt über Lautsprecher. Ausgehend von der statischen Übersprechkompensation für einen Punkt wurde ein System realisiert, welches die Kompensationsfilter in Abhängigkeit der momentanen Kopfposition in Echtzeit berechnet. Dies ermöglicht die korrekte Wiedergabe innerhalb eines größeren Raumbereichs. Die Evaluierung des Verfahrens hat eine sehr gute Kanaltrennung für die erste Wellenfront gezeigt, die in subjektiven Untersuchungen in nicht reflexionsfreier Umgebung bestätigt werden konnte.

Das mit den beschriebenen Techniken realisierte Audio System wurde in das am

Rechen- und Kommunikationszentrum der RWTH vorhandene VR System (CAVE-like environment) integriert. Das Gesamtsystem erlaubt die Erzeugung komplexer multi-modaler virtueller Umgebungen und wird bereits in einer Reihe interdiszipli-närer Forschungsstudien verwendet.

Appendix

Glossary

Abbreviations

Abbreviation	Description
CCC	Center for Computing and Communication
CTC	Crosstalk Cancellation
FFT	Fast Fourier Transform
FTD	Frontal Time Delay
HMD	Head-Mounted Display
HRIR	Head-Related Impulse Response
HRTF	Head-Related Transfer Function
IFFT	Inverse Fast Fourier Transform
ILD	Interaural Level Difference
ITD	Interaural Time Difference
ITA	Institute of Technical Acoustics
LCD	Liquid Crystal Display
LTI	Linear Time Invariant
RCS	Real Coordinate System
SNR	Sound to Noise Ratio
VBAP	Vector Base Amplitude Panning
ViSTA	Virtual Reality for Scientific Technical Applications
VR	Virtual Reality
WCS	World Coordinate System
WFS	Wave-Field Synthesis

Directivity Plots

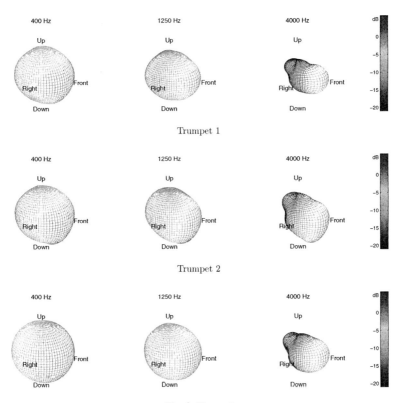

142

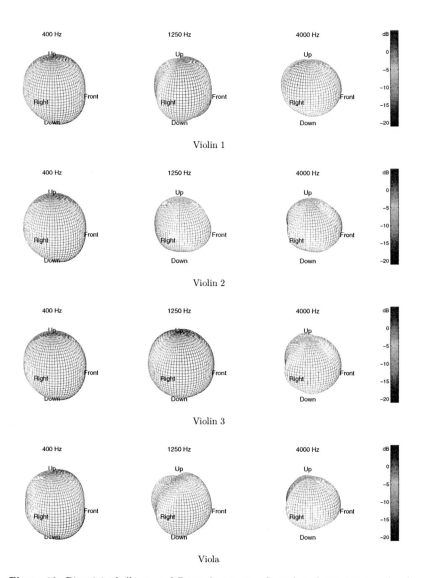

Figure 12: Directivity balloons at different frequencies. Spatial resolution is interpolated to 5 degree.

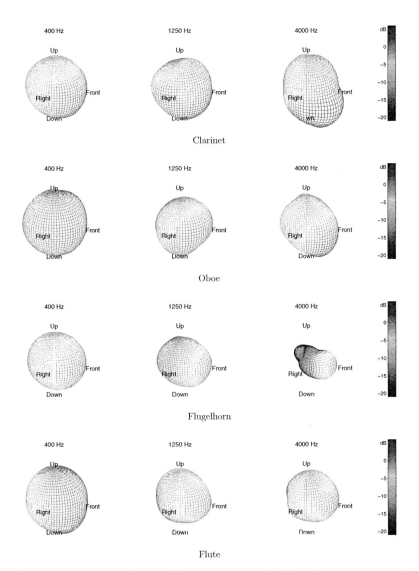

Figure 13: Directivity balloons at different frequencies. Spatial resolution is interpolated to 5 degree.

Curriculum vitae

Personal Data

	Tobias Lentz
29.08.1971	born in Rheydt, Germany
	married, two children
	E-mail: tobias.lentz@gmx.de

Education

08/1978 - 06/1982	Primary School, "Montessori Grundschule", Mönchengladbach
08/1982 - 06/1991	Secondary School, "Mathematisch-Naturwissenschaftliches Gymnasium", Mönchengladbach
09/1991 - 11/1992	Civil service, "Werkstattgruppe im Zentrum für Körperbehinderte", Mönchengladbach

Course of Studies

10/1993 - 10/2001	Study of Electrical Engineering at RWTH Aachen University
04/2001 - 10/2001	Diploma Thesis at the Institute of Technical Acoustics, RWTH Aachen University
11/2001 - 04/2007	Ph.D. at the Institute of Technical Acoustics, RWTH Aachen University

Employments

04/2000 - 06/2000	Internship at BOSE Corporation, Automotive Systems Division, Esslingen, Germany and Framingham, USA
11/2001 - 04/2007	Research Assistant at the Institute of Technical Acoustics, RWTH Aachen University
since 05/2007	Employe at HEAD acoustics GmbH, Herzogenrath-Kohlscheid, Germany

Danksagung

Viele Personen waren an der Entstehung dieser Arbeit beteiligt, bei denen ich mich an dieser Stelle ganz herzlich bedanken möchte.

Professor Dr. rer. nat. Michael Vorländer danke ich ganz, ganz herzlich für die Betreuung dieser Arbeit, für die fachliche Unterstützung und Begleitung. Besonders möchte ich mich auch für die Schaffung der kreativen Umgebung im Institut für Technische Akustik bedanken, in der es immer möglich war auf dem kurzen Dienstweg Fragen und Ideen zu diskutieren.

Professor Christian Bischof, Ph.D. möchte ich für die Übernahme des Koreferates und die Unterstützung des gesamten DFG-Projektes ganz herzlich danken.

Den Werkstätten des Instituts sei für die großartige Unterstützung gedankt. Ohne die zahlreichen Ideen und die engagierte Umsetzung wären die vielen Versuchs- und Messaufbauten nicht möglich gewesen. Meinen Diplomanden, Studienarbeitern und studentischen Hilskräften Christian Buchcik, Hilmar Demuth, Nils Frohmüller, Robert Jost, Jan Köhler, Ben Marpe, Steven Phee, Alexander Pohl, Christian Renner, Dirk Schröder, Karl Slenczka, Aulis Telle, Frank Wefers und Martin Zarzycki danke ich sehr für den Einsatz bei der Bearbeitung von etlichen Teilaspekten der Arbeit. Dr.-Ing. Gottfried Behler danke ich für die zahlreichen und fruchtbaren Diskussionen, sowie für die aufmunternde Art die Dinge wieder zurechtzurücken, wenn das Gedanken-Chaos zu groß wurde. Allen Mitarbeitern des Instituts für Technische Akustik möchte ich für die tolle Zeit, den ein und anderen verrückten Abend und die rundum gute Atmosphäre danken. Dem ITA danke ich für die Kaffeeecke. Danken möchte ich auch Stephanie Heikamp für das Korrekturlesen.

Da diese Arbeit aus dem DFG Gemeinschaftsprojekt „Der virtuelle Kopfhörer" des Instituts für Technische Akustik und des Rechen- und Kommunikationszentrums hervorgegangen ist, möchte ich an dieser Stelle auch allen, die außerhalb des ITA an diesem Projekt beteiligt waren, danken. Ohne die Infrastruktur (besonders die CAVE) des VR-Labors im Rechen- und Kommunikationszentrum wäre die Arbeit sicherlich nicht in diesem Umfang möglich gewesen. Für die Schaffung dieser Voraussetzungen und für viele wertvolle Diskussionen möchte ich mich bei Dr. rer. nat. Torsten Kuhlen ganz herzlich bedanken. Ebenso danke ich allen Mitarbeitern der VR-Gruppe im Rechenzentrum für die nette Zusammenarbeit und Unterstützung.

Ganz besonders möchte ich mich bei Ingo Assenmacher, mit dem ich gemeinsam das DFG-Projekt bearbeiten durfte, für die grandiose Zusammenarbeit, und auch für die vielen fachlichen und persönlichen Diskussionen bei den sogenannten „Arbeitstreffen" bedanken. Es war für mich eine große Freude in diesem Zweierteam zu arbeiten.

Schließlich möchte ich mich ganz herzlich bei meiner Familie bedanken, die mich besonders während der heißen Schlussphase der Promotion sehr unterstützt hat.

Bibliography

[AAD01]   V. R. Algazi, C. Avendano and R. O. Duda, *Estimation of a Spherical-Head Model from Anthropometry*, Journal of the Auidio Engineering Society **49** (2001), no. 6, 472–479.

[AB79]   J. Allen and D. A. Berkley, *Image Method for Efficiently Simulating Small-Room Acoustics*, The Journal of the Acoustical Society of America **65(4)** (1979), 943–950.

[AB94]   R. Azuma and G. Bishop, *Improving static and dynamic registration in an optical see-through HMD*, SIGGRAPH '94: Proceedings of the 21st annual conference on Computer graphics and interactive techniques, New York, NY, USA, ACM Press, 1994, pp. 197–204.

[AB95]   R. Azuma and G. Bishop, *A frequency-domain analysis of head-motion prediction*, SIGGRAPH '95: Proceedings of the 22nd annual conference on Computer graphics and interactive techniques, New York, NY, USA, ACM Press, 1995, pp. 401–408.

[AKL05]   I. Assenmacher, T. Kuhlen and T. Lentz, *Binaural Acoustics for CAVE-like Environments without Headphones*, Eurographics Symposium on Virtual Environments, Eurographics ACM SIGGRAPH, Aalborg, Denmark, 2005.

[ALK06]   I. Assenmacher, T. Lentz and T. Kuhlen, *Binaural Synthesis for CAVE-like Environments without Headphones*, Proceedings of the 16th World Congress on Ergonomics, IEA2006, Maastricht, Netherlands, 2006.

[Ama00]   J. Amate, *Investigation of parameters for dynamic crosstalk cancellation*, Diplomarbeit, Institute of Technical Acoustics, RWTH Aachen University, Germany, 2000.

[ART04]   ART, *A.R.T. tracking systems, http://www.ar-tracking.de*, A.R.T. GmbH, 2004.

[AS63]   B. S. Atal and M. R. Schröder, *Apparent sound source translator*, Tech. report, US Patent 3,236,949, February 23 1963.

[Asc]   Ascension, *Flock of Birds - Installation and Operation Guide, http://www.ascension-tech.com*.

[Bau63]   B. B. Bauer, *Stereophonic Earphones and Binaural Loudspeakers*, Journal of the Audio Engineering Society **9** (1963), no. 2, 148–151.

[BC96]   J. Bauck and D. H. Cooper, *Generalization Transaural Stereo and Applications*, Journal of the Audio Engineering Society **44** (1996), no. 9, 683–705.

[BDR99]   D. S. Brungart, N. I. Durlach and W. M. Rabinowitzb, *Auditory localization of nearby sources. II. Localization of a broadband source*, Journal of the Acoustical Society of America **106** (1999), 1956–1968.

[Beg91]   D. R. Begault, *Challenges to the Successful Implementation of 3-D Sound*, Journal of the Audio Engineering Society **39** (1991), no. 11, 864–870.

[Bla97]   J. Blauert, *Spatial Hearing: The psychophysics of Human Sound Localization, Revised Edition*, revised edition ed., MIT Press, Cambridge MA, 1997.

[Bla05]   J. Blauert (ed.), *Communication Acoustics*, Springer, 2005.

[BR99]   D. S. Brungart and W. M. Rabinowitzb, *Auditory localization of nearby sources. Head-related transfer functions*, Journal of the Acoustical Society of America **106** (1999), 1465–1479.

[BSM+04] D. S. Brungart, D. D. Simpson, R. L. McKinley, A. J. Kordik, R. C. Dall-
man and D. A. Ovenshire, *The Interaction Between Head-Tracker Latency, Source Duration, and Response TIme in the Localization of Virtual Sound Sources*, Proceedings of ICAD 04 - Tenth Meeting of the
International Conference on Auditory Display, Sidney, Australia, 2004.

[BT05] T. Brookes and C. Treble, *The Effect of Non-Symmetrical Left/Right
Recording Pinnae on the Perceived Externalisation of Binaural Record-
ings.*, Proceedings of the 118th Audio Engineering Society Convention
Barcelona, Spain, Preprint 6439, 2005.

[Buc04] C. Buchcik, *Die Grenzen des Sweet Spots bei der statischen Übe-
sprechkompensation*, Studienarbeit, Institute of Technical Acoustics,
RWTH Aachen University, Germany, 2004.

[BVdV92] A. Berkhout, P. Vogel and D. de Vries, *Use of Wave Field Synthesis for
Natural Reinforced Sound*, Proceedings of the Audio Engineering Society
Convention 92, no. Preprint 3299, 1992.

[CHV02] L. Chai, W. A. Hoff and T. Vincent, *Three-dimensional motion and structure estimation using inertial sensors and computer vision for augmented
reality*, Presence: Teleoper. Virtual Environ. **11** (2002), no. 5, 474–492.

[CNSD93] C. Cruz-Neira, D. J. Sandin and T. A. DeFanti, *Surround-screen
projection-based virtual reality: the design and implementation of the
CAVE*, SIGGRAPH '93: Proceedings of the 20th annual conference on
Computer graphics and interactive techniques, New York, NY, USA,
ACM Press, 1993, pp. 135–142.

[DC06] R. M. Dizon and H. S. Colburn, *The influence of spectral, temporal, and
interaural stimulus variations on the precedence effect*, The Journal of
the Acoustical Society of America **119** (2006), no. 5, 2947–2964.

[Fro05] N. Frohmüller, *Untersuchungen zur Kanaltrennung bei der dynamischen
Übersprechkompensation*, Diplomarbeit, Institute of Technical Acoustics,
RWTH Aachen University, Germany, 2005.

[Gar68]   M. B. Gardner, *Historical Background of the Haas and or Precedence Effect*, Journal of the Acoustical Society of America **43** (1968), 1243–1248.

[Gar95]   W. G. Gardner, *Efficient convolution without input-output delay*, Journal of the Audio Engineering Society **43 No. 3** (1995), 127 – 136.

[Gar97]   W. G. Gardner, *3-D audio using loudspeakers*, Ph. D. thesis, MIT Media Lab, Massachusetts Institute of Technology, 1997.

[Gar02]   G. García, *Optimal Filter Partition for Efficient Convolution with Short Input/Output Delay*, Proceedings of the 113th Audio Engineering Society Convention, Los Angeles, USA, 2002.

[HM91]   D. Hammershoø and H. Møller, *Ree-field sound transmission to the external ear; a model and some measurements*, Fortschritte der Akustik - DAGA, Bochum, Germany, 1991, pp. 473–476.

[HP04]   K. Hartung and R. Pellegrini, *Vergleich verschiedener Interpolationsalgorithmen für Außenohrübertragungsfunktionen in dynamischen auditiven Umgebungen*, Fortschritte der Akustik - DAGA, Oldenburg, Germany, 2004.

[KC98]   A. Kulkarni and H. S. Colburn, *Role of spectral detail in sound-source localization*, Nature **396** (1998), no. 24, 747–749.

[KC00]   A. Kulkarni and H. S. Colburn, *Variability in the characterization of the headphone transfer-function*, The Journal of the Acoustical Society of America **107** (2000), no. 2, 1071–1074.

[KC05]   S. M. Kim and W. Choi, *On the externalization of virtual sound images in headphone reproduction: A Wiener filter approach*, The Journal of the Acoustical Society of America **117** (2005), no. 6, 3657–3665.

[KS93]   J. Köring and A. Schmitz, *Simplifying Cancellation of Cross-Talk for Playback of Head-Related Recordings in a Two-Speaker System*, ACUSTICA **79** (1993), 221–232.

[Kut95]   H. Kuttruff, *A simple iteration scheme for the computation of decay constants in enclosures with diffusely reflecting boundaries*, The Journal of the Acoustical Society of America **98(1)** (1995), 288–293.

[Kut00]   H. Kuttruff, *Room Acoustics, Fourth Edition*, Elsevier Science Publisher, 2000.

[KW03]   S. M. Kim and S. Wang, *A Wiener filter approach to the binaural reproduction of stereo sound*, Journal of the Acoustical Society of America **114** (2003), no. 6, 3179–3188.

[LAS05]   T. Lentz, I. Assenmacher and J. Sokoll, *Performance of Spatial Audio Using Dynamic Cross-Talk Cancellation*, Proceedings of the 119th Audio Engineering Society Convention New York, USA, Preprint 6541, 2005.

[LAVK06]   T. Lentz, I. Assenmacher, M. Vorländer and T. Kuhlen, *Precise Near-to-Head Acoustics with Binaural Synthesis*, Journal of Virtual Reality and Broadcasting **3** (2006), no. 2, Online Journal, `urn:nbn:de:0009-6-5890`, ISSN 1860-2037.

[Len03]   T. Lentz, *Untersuchungen zum Einfluss von Reflexionen bei der Übersprechkompensation*, Fortschritte der Akustik - DAGA, Aachen, Germany, 2003.

[Len06]   T. Lentz, *Dynamic Crosstalk Cancellation for Binaural Synthesis in Virtual Reality Environments*, Journal of the Audio Engineering Society **54** (2006), no. 4, 283–294.

[Len07]   T. Lentz, *Near-Field Hrtfs*, Fortschritte der Akustik - DAGA, Stuttgart, Germany, 2007.

[LHS99]   T. Lokki, J. Hiipakka and L. Savioja, *Immersive 3-D Sound Reproduction in a Virtual Room*, Proceedings of the AES 16th International Conference: Spatial Sound Reproduction, 1999.

[LS06]   T. Lentz and K. Slenczka, *Simulation of Natural Sound Sources*, Fortschritte der Akustik - DAGA, Braunschweig, Germany, 2006.

[LSVA07]   T. Lentz, D. Schröder, M. Vorländer and I. Assenmacher, *Virtual Re-
           ality System with Integrated Sound Field Simulation and Reproduction*,
           EURASIP Journal on Advances in Signal Processing, Special Issue on
           Spatial Sound and Virtual Acoustics (2007), Article ID 70540, 19 pages.

[LVJ03]   J. J. La Viola Jr., *A Testbed for Studying and Choosing Predictive Track-
           ing Algorithms in Virtual Environments*, 7. International Immersive Pro-
           jection Technologies Workshop, 9. Eurographics Workshop on Virtual
           Environments, 2003.

[Mey99]   J. Meyer, *Akustik und Musikalische Aufführungspraxis*, 4 ed., Verlag Er-
           win Brochinsky, 1999.

[MK99]   P. Mouchtaris, A.and Reveliotis and C. Kyriakakis, *Non-minimum Phase
           Inverse Filter Methods for Immersive Audio Rendering, Phoenix*, Pro-
           ceedings IEEE International Conference on Acoustics, Speech and Signal
           Processing (ICASSP), 1999.

[Møl89]   H. Møller, *Reproduction of artificial head recordings through loudspeak-
           ers*, Journal of the Audio Engineering Society **37** (1989), no. 1/2, 30–33.

[Møl92]   H. Møller, *Fundamentals of Binaural Technology*, Applied Acoustics **36**
           (1992), 171–218.

[MPO⁺00]   P. Minnaar, J. Plogsties, S. K. Olesen, F. Christensen and H. Møller,
           The Interaural Time Difference in Binaural Synthesis, Proceedings of
           the 108th Audio Engineering Society Convention, Paris, France, 2000.

[NMS92]   G. Neu, E. Mommertz and A. Schmitz, *Untersuchungen zur rich-
           tungstreuen Schallwiedergabe bei Darbietung von kopfbezogenen Aufnah-
           men über zwei Lautsprecher I*, ACUSTICA **77** (1992), 183–192.

[NSG02]   M. Naef, O. Staadt and M. Gross, *Spatialized Audio Rendering for Im-
           mersive Virtual Environments*, Proceedings of the ACM symposium on
           Virtual reality software and technology, Hong Kong, China, 2002, pp. 65
           – 72.

[ORC⁺02] F. Otondo, J. Rinde, R. Caussé, O. Misdariis and P. D. la Cuadra, *Directivity of musical instruments in a real performance situation*, Proceedings of the International Symposium on Musical Acoustics, Mexico city, Mexico, 2002, pp. 312–318.

[Phe02] S. Phee, *Near-Field Measurements of Head-Related Transfer Functions*, Studienarbeit, Institute of Technical Acoustics, RWTH Aachen University, Germany, 2002.

[PS90] D. R. Perrott and K. Saberi, *Minimum audible angle thresholds for sources varying in both elevation and azimuth*, The Journal of the Acoustical Society of America **87** (1990), no. 4, 1728–1731.

[RB68] S. K. Roffler and R. A. Butler, *Factors That Influence the Localization of Sound in the Vertical Plane*, Journal of the Acoustical Society of America **43** (1968), no. 6, 1255–1259.

[RF95] C. Ryan and D. Furlong, *Effects of Headphone Placement on Headphone Equalization for Binaural Reproduction*, Proceedings of the 98th Audio Engineering Society Convention, Paris, France, 1995.

[RME] RME, *Manual Hammerfall System, Multiface, http://www.rme-audio.de.*

[RNRT02] J. Rose, P. Nelson, B. Rafaely and T. Takeuchi, *Sweet spot size of virtual acoustic imaging systems at asymmetric listener locations*, The Journal of the Acoustical Society of America **112** (2002), no. 5, 1992–2002.

[ROC04] J. H. Rindel, F. Otondo and C. L. Christensen, *Sound Source Representation for Auralization*, Proceedings of the International Symposium on Room Acoustics: Design and Science, 2004.

[RV03] G. Romanenko and M. Vorländer, *Employment of spherical wave reflection coefficient in room acoustics*, IoA Symposium Surface Acoustics, Salford, U.K., 2003.

[SBGS69]   R. Shumacker, R. Brand, M. Gilliland and W. Sharp, *Study for Applying Computer-Generated Images to Visual Simulations*, Report AFHRL-TR-69-14, U.S. Air Force Human Resources Laboratory, 1969.

[Sch93]   A. Schmitz, *Naturgetreue Wiedergabe kopfbezogener Schallaufnahmen über zwei Lautsprecher mit Hilfe eines Übersprechkompensators.*, Ph. D. thesis, Institute of Technical Acoustics, RWTH Aachen University, Germany, 1993.

[SDV07]   D. Schröder, P. Dross and M. Vorländer, *A Fast Reverberation Estimator for Virtual Environments*, AES 30th International Conference, Saariselkä, Finland, 2007.

[SHLV99]   L. Savioja, J. Huopaniemi, T. Lokki and R. Väänänen, *Creating Interactive Virtual Acoustic Environments*, Journal of the Audio Engineering Society **47** (1999), no. 9, 675–705.

[SL06]   D. Schröder and T. Lentz, *Real-Time Processing of Image Sources Using Binary Space Partitioning*, Journal of the Audio Engineering Society **54** (2006), no. 7/8, 604–619.

[SLA06]   J. Sokoll, T. Lentz and I. Assenmacher, *Evaluierung des virtuellen Kopfhörers für die dynamische binaurale Synthese*, Fortschritte der Akustik - DAGA, Braunschweig, 2006.

[Sle04]   K. Slenczka, *Simulation of Natural Instruments for Binaural Synthesis*, Diplomarbeit, Institute of Technical Acoustics, RWTH Aachen University, Germany, December 2004.

[SN97]   J. S. Suh and P. A. Nelson, *Measurements of transient response of room and comparison with geometrical acoustic models*, The Journal of the Acoustical Society of America **105(4)** (1997), 2304–2317.

[Ste04]   Steinberg, *ASIO 2.0 Audio Streaming Input Output Development Kit*, 2004.

[TNKH97]   T. Takeuchi, P. Nelson, O. Kirkeby and H. Hamada, *The effects of re-flections on the performance of virtual acoustic imaging systems*, Proceedings Active '97, Budapest, Hungary, 1997.

[TR67]   W. R. Thurlow and P. S. Runge, *Effect of Induced Head Movements on Localization of Direction of Sounds*, Journal of the Acoustical Society of America **42** (1967), 480–488.

[VdPK00]   S. Van de Par and A. Kohlrausch, *Sensitivity to Auditory-Visual Asynchrony and to Jitter in Auditory-Visual Timing*, Human Vision and Electronic Imaging V, Proceedings of the SPIE (B. E. Rogowitz and T. N. Pappas, eds.), vol. 3959, 2000, pp. 234–242.

[Vor00]   M. Vorländer, *Acoustic load on the ear caused by headphones*, Journal of the Acoustical Society of America **107** (2000), no. 4, 2082–2088.

[WAKW93]   E. Wenzel, M. Arruda, D. J. Kistler and F. Wightman, *Localisation using nonindividualized head-related transfer functions*, Journal of the Acoustical Society of America **94 (1)** (1993), 111–123.

[War01]   D. B. Ward, *On the performance of acoustic crosstalk cancellation in a reverberant environment*, Journal of the Acoustical Society of America **110** (2001), no. 2, 1195–1198.

[WD07]   I. Witew and P. Dietrich, *Assessment of the Uncertainty in Room Acoustical Measurements*, 19th ICA, Madrid, Spain, 2007.

[Wef07]   F. Wefers, *Optimizing segmented realtime convolution*, Diplomarbeit, Center for Computing and Communication, RWTH Aachen University, Germany, September 2007.

[Wit04]   I. Witew, *Spatial variation of lateral measures in different concert halls*, 18th ICA, Kyoto, Japan, vol. 4, 2004, p. 2949.

[WK99]   F. L. Wightman and D. J. Kistler, *Resolution of front-back ambiguity in spatial hearing by listener and source movement*, Journal of the Acoustical Society of America **105** (1999), no. 5, 2841–2853.

[WO95]   J. R. Wu and M. Ouhyoung, *A 3D tracking experiment on latency and its compensation methods in virtual environments*, UIST '95: Proceedings of the 8th annual ACM symposium on User interface and software technology, New York, NY, USA, ACM Press, 1995, pp. 41–49.